THE HERDSMAN'S BOOK

THE HERDSMAN'S BOOK

J. MALCOLM STANSFIELD

BSc, POST-GRAD DIPLOMA FARM MANAGEMENT,
FBIM, FRAgS

Farming Press

First published 1983 as *The New Herdsman's Book*
Second Edition 1991 retitled as *The Herdsman's Book*

A catalogue record for this book is available from the British Library

ISBN 0 85236 215 3

Cover photo of a herdsman driving Friesian cows
along a country track at Sonning Farm in Berkshire
supplied by Holt Studios Photography Library, Hungerford, Berkshire

Published by Farming Press Books
4 Friars Courtyard, 30–32 Princes Street
Ipswich IP1 1RJ, United Kingdom

Distributed in North America
by Diamond Farm Enterprises,
Box 537, Alexandria Bay, NY 13607, USA

Phototypeset by Galleon Photosetting, Ipswich
Printed and bound in Great Britain by
Butler and Tanner, Frome, Somerset

CONTENTS

A colour section follows page 148

PREFACE TO THE SECOND EDITION

Since the first edition of this book was published, a number of important developments have taken place in the cattle industry, all of which affect the role of the herdsman in one way or another.

The introduction of milk quotas within the European Community in 1984 is perhaps the best example, necessitating the increased need to monitor and control milk output and to adjust feeding and management so as to optimise margins per litre.

The seasonality payments for milk have markedly changed, increasing the value of summer milk, so requiring more input to the planning of preferable calving dates, especially of heifers. July and August calving may be ideal for the cows, if adequate feed is available, and good for the herd owner's bank account, but for a herdsman with a young family used to taking a summer break in the school holidays, difficulties obviously arise.

'Green' issues have in recent years become more prominent in agriculture as a whole, with beef and particularly dairy farming being no exception. Increasing consideration has to be given to such topics as animal welfare, silage effluent, slurry spreading and COSHH regulations, to name but a few. Herdsmen, like all members of the industry, have a major role to play in improving the public image of farming and hopefully restoring the confidence of our customers in meat, milk and dairy products.

Even the confidence of some herdsmen in the future prospects for the industry has been shaken by, for example, the problems of BSE (particularly its media coverage) and by the increasing incidence of leptospirosis with its risk to personal health.

This revised edition 'takes on board' all of these issues, hopefully puts them into context, and, as with the original text, aims to help those working with cattle to get satisfaction from coping with their many problems.

I am extremely grateful to Michael Collis and his colleagues at Sonning Farm for comments on the text and invaluable assistance with the production of photographs. My particular thanks are due to Ian Maclean who again took many of the photographs, to my brother, Roger, for his input to the section on lameness and foot care, and to Mrs Sheila Smith who typed (sorry–wordprocessed this time) the manuscript.

Finally, I wish to thank Roger Smith of Farming Press for his continued help and support which ensured the arrival 'on the scene' of this revised text.

MALCOLM STANSFIELD
November, 1990

PREFACE TO THE FIRST EDITION

This book appears, not because of any ambitions to become an author, but due to a concern that people working with cattle should do a good job, make the most of their changing situations and obtain maximum satisfaction from their work. Even my school reports would endorse this as an appropriate objective, especially those from my English Master who stated with some regularity, 'If this pupil took half as much interest in his studies as he does in his farming he would go far'! He would smile today if he only knew how many drafts have been necessary to produce this text.

My interest in cattle was aroused at an early age on the local dairy farm of Brewis Cant, a friend of the family. He typified the first-class stockmen of the Yorkshire Dales but perhaps was exceptional in his patience, understanding and ability to get the most out of youngsters, as well as cows. I was taught the correct ways of undertaking husbandry tasks: that the cows come first but that every farm resource is important and needs care, whether it be a field of baled hay or the yard broom. The factors which determine success in dairying were clearly demonstrated, such as: good feeding, regular breeding and prompt attention to a sick animal. Most importantly, I was taught to *think* before acting and to complete every task to the best of my ability, especially when cattle and other people were involved.

This book appears, therefore, because I consider that so many of those factors continue to be critical to success. The modern herdsman may well be operating with cattle of higher potential and with a wide range of sophisticated aids, but he or she is still working with the same basic resources; and so as well as a knowledge of modern technology, they need the right attitude to the job.

It is a great privilege to have my name on a publication which in many ways follows Kenneth Russell's original *Herdsman's Book*. I first met Ken when I was visiting Askham Bryan as an undergraduate student and in later years I heard him speak to farmers on many occasions. He was a truly great man and his words on stockmanship were 'music to my ears'. On the day he arrived in the farmyard at Sonning for a walk through 'my' herd and a discussion on the management of large dairies,

I thought I had then achieved all that was possible in dairying.

Like his books, this one is written in the hope that it will be of help to practical herdsmen, to their employers, as well as to all students of dairying. Like Ken, too, I admire good cows and good herdsmen—and the one deserves the other.

I am grateful to many colleagues and friends in the industry for their help, advice and supply of information. My very special thanks are due to Tony Giles and to Bill Milligan for their invaluable encouragement and also for reading and suggesting amendments to major parts of the manuscript.

I would also like to thank Mrs Frances Green and Mrs Sheila Smith for typing the several drafts and for their general help in the preparation.

To the many local farmers and herdsmen and especially to the staff of Sonning Farm who so willingly co-operated in the production of photographs, I am also most grateful. My thanks are also due to Ian Maclean who took the majority of the pictures, patiently protecting his equipment from slurry and the licking of friendly cows. The remainder were kindly supplied by BOCM Silcock Ltd, *Dairy Farmer*, Tim Bryce, Tony Quick and my son John.

Finally, my special thanks to my long-suffering wife, Mary, who began to believe it would never be finished—and to Roger Smith of Farming Press for his constant help and advice in order to ensure that it was.

INTRODUCTION

With the continuing trend towards mechanisation and automation of livestock farms, some may wonder if the days of the stockman* are numbered. It is a fact that on many dairy farms electronically controlled equipment has replaced the herdsman for such tasks as concentrate feeding and cluster removal. On a few units, automation of milk recording, teat disinfection and cow weighing is in operation, with results usually more accurate than when undertaken by stockmen. In the opinion of this writer, however, it will be a remarkable day when the dairy robot is able to cope with a difficult calving or to persuade a dozy, new-born calf to suckle its dam!

In the future, many aspects of the herdsman's job will no doubt continue to change. He may further reduce his physical work and spend more time supervising equipment, but he will retain, and even expand, the time spent on the vital task of caring for the welfare of his cattle.

Change is not a new experience for most herdsmen, at least for those working with cattle during the last fifty years. The average dairy herd has more than doubled in size, breeds of both dairy and beef cattle have come and gone and machinery has progressively replaced much of the hand work involved in feeding, milking and manure handling. Specialisation has become a feature of many farms, with herds being increased and systems intensified to cover the rising background costs of such items as labour and machinery.

Not all cattle enterprises have changed quite so dramatically. Although very few animals are now hand milked, many smaller farms still operate with cowshed housing and use minimal mechanisation. The change in the role of beef herdsmen has in general been less marked than with dairying, especially for those

* Or woman. It would be invidious to suggest that one sex was better at caring for stock than the other. A single gender will be used to avoid unnecessary repetition.

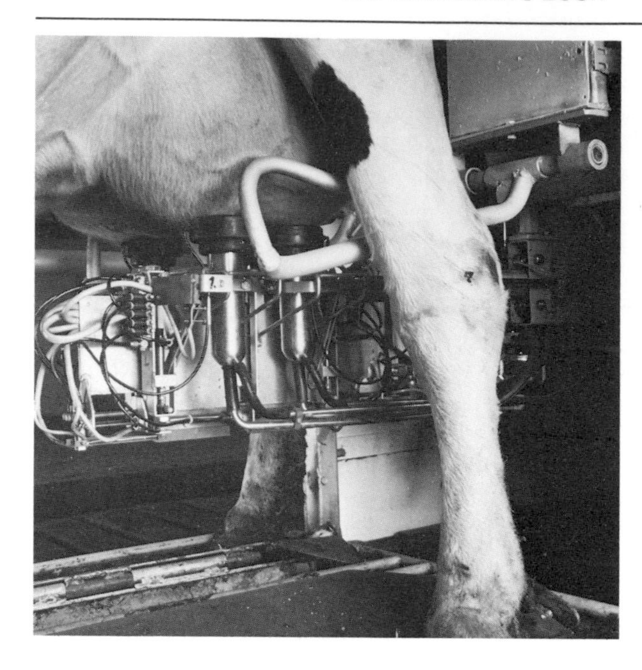

A robotic milking machine. Development of such equipment is being undertaken in many countries by commercial companies and research institutes.

involved with suckler herds. Many such herds are carried on holdings where the climatic and physical environment limits the economic intensification of the enterprise. Others are integrated as low-cost units utilising by-products on arable farms. However, on many beef-fattening units stockmen are involved in technology as high as that in dairying with sophisticated feeding and management systems.

The detailed tasks that make up the job of each stockman do therefore vary considerably from farm to farm but there are some which are common to all situations. An attempt has been made in the following chapters to discuss all the most important tasks and considerations so that the book should be of interest to anyone involved with cattle enterprises regardless of their types or sizes.

It is assumed that all herdsmen are keen and dedicated to their cattle, being fully aware of the satisfaction that can be obtained from a job well done. One of the great pleasures of working with stock is to see the results of one's efforts, sometimes almost immediately, as with a cow responding to injection for grass staggers. The job also has its share of frustrations and disappointments—sometimes even more than its share!

The objective of this book is to foster a balanced approach to the job and to encourage herdsmen to think and plan, but above all to carry out timely and appropriate action. It is hoped that it will encourage herdsmen to correct their weaknesses (which everyone has) and to build on strengths by developing skills (which everyone also has). Tackling a job with understanding and confidence leads to even greater satisfaction and contributes much to improvements in the physical and financial performance of the enterprise.

Financial aspects of the herd have traditionally been of minor interest to herdsmen but this is a situation which is now rapidly changing. In most high-performance enterprises, farmers and managers involve their herdsmen in many of the business tasks such as setting targets and discussing results, in a way which usually provides a considerable stimulus to motivation.

On the numerous one-man units, many of which are family farms, the herd owner is also stockman so that he has even more demands on his time. Routine tasks such as feeding naturally take priority, but when problems arise such as a sick cow the business aspects and office work just have to wait.

In all herd situations there is an important need for effective communications between herdsmen and the various people who

A large herd ready for milking. The centrally mounted gate in this rotary collecting yard moves around behind the cows, encouraging them to enter the parlour.

are involved in the operation of the enterprise. With small herds, apart from members of the family, this may involve only the veterinary surgeon, AI inseminator and perhaps a relief milker. In large herds it will involve other herdsmen and often tractor drivers, staff from the farm's other enterprises as well as a wide range of off-farm people such as advisors and representatives.

Where there are large herds with several herdsmen involved a good team spirit is essential to success. Each member needs to enjoy working with other people and to understand the interests and motivations of his colleagues. Sharing the less pleasant aspects of the work such as slurry scraping always helps. There is a considerable challenge for employers and managers in these conditions to set up and control an effective system of operation. Team members need to be selected with suitable personalities for this rather special situation. Researchers have shown that many herdsmen prefer to work alone and to take full responsibility for the day-to-day operation of a herd. This type of herdsman apparently tends to be an introvert and may be keen on such hobbies as gardening. He is not concerned about the long working day which offers few opportunities to get away from the farm. However, herdsmen operating in teams commonly work shifts so that at intervals during the rota they have useful blocks of time which can be spent with the family (at meal times) or participating in leisure activities off the farm.

It appears, therefore, that despite the arrival of modern technology on the cattle farm there is still an important job to be done by herdsmen. The successful ones will increasingly be those with a wide range of skills, a detailed knowledge of what 'makes a cow tick' and the ability to carry out the right job at the right time. They will also have an increasingly important role in the general management of cattle enterprises by being 'spokesmen' for the cattle in making appropriate representations when decisions need to be taken about such items as building design, seed mixtures and diet formulation.

It is the author's hope that this book will help stockmen to undertake this changing role with confidence.

CHAPTER 1

GENERAL STOCKMANSHIP

There is no doubt some truth in the saying that a good stockman is born and not trained, as many of his essential qualities like patience, reliability and a sense of caring are very much part of his character. However, other equally essential qualities such as his skills of observation, handling of stock and the undertaking of his many husbandry tasks are gained by experience and training.

Stockmanship is not easy to define. It is well recognised that the person with such talents has 'a way with stock'. There exists a special relationship between man and animal: each understands the other and they even appear to be able to communicate with each other. Watching a skilled herdsman amongst his cattle, one can see stockmanship in action. He operates with confidence, speaks to the animals in a reassuring tone of voice and often touches them, so that they show no signs of fear and seldom back away.

The results of his sense of caring can be seen in the general health of the stock, and their environment: the bloom of the coat, clean, dry cubicle beds, the absence of overgrown feet.

Traditionally, the herdsman had to know every animal in the herd so as to be able to manage each one as an individual. The situation in many of today's larger herds, with several herdsmen involved and with considerable automation, is somewhat different. Nevertheless, any animal seen to be behaving in an abnormal or unexpected way has to be identified immediately and the records checked before any subsequent action can be taken.

Observation

Careful and critical observation, not only of individual animals but also of the group as a whole, is therefore an essential feature of good stockmanship. It involves most of the senses—sight, touch, smell, hearing—but rarely taste! Careful observation can

Young herdsmen and women taking part in a training session. The particular task under instruction is the dosing of a young heifer.

A demonstration of stockmanship. The good herdsman operates with confidence, speaks to his animals in a reassuring tone of voice and often touches them, so that they show no signs of fear and seldom back away.

take quite a lot of energy so that at times of heavy physical work there is a danger of becoming over-tired and observing less accurately.

Skilled observation comes with experience but it can be greatly enhanced by training, especially by working alongside a skilled stockman who at every opportunity points out to the trainee any irregularities and confirms the normal situations. Apart from a few hours in the middle of the night when cows are all resting, there is usually something interesting to observe, and even then, a few cows at peak yield may be eating.

It is not only the animals which have to be observed; attention must also be given to the numerous aspects of the environment which influence animal behaviour and performance. Excessively wet bedding may, for example, indicate a fracture in an underground water pipe or a blocked drain rather than any problem with the straw or with the animals. Attention must even be given to colleagues, especially when one is responsible for *their* work. A person under the weather may not be aware that his performance is slipping but if he is not fully alert he is more likely to be involved in an accident.

Animal Behaviour

A herdsman will find it easier to observe and handle the animal effectively if he is aware of the normal behavioural pattern of a herd of cattle. A number of factors influence behaviour, one of the most obvious being the oestrous cycle in cows and mature heifers, which will be discussed in Chapter 11.

The social ranking within a herd has a marked effect upon behaviour. Some animals are more aggressive than others and there is an aggressive ranking or 'bunt order' within a group, with the boss animals at the top and timid ones at the bottom. The larger, older cows in a milking group tend to be of high rank and first-lactation heifers and recently calved animals of lower rank. This factor is particularly important in the management of a herd which is loose-housed or self-fed silage. Boss cows will stand at strategic places such as the silage face, keeping subordinate ones away so that they fail to get adequate feed. Tombstone barriers or feed fences fitted with diagonal rails reduce such bullying, as boss animals are prevented from moving quickly back from the feeding face. Subdivision of a large group into sub-groups is a management practice which also reduces bullying. A milking cow in a group of no more than 80 seems to be able to readily recognise the ones higher in the 'bunt order' and so moves to one side to let them pass. In groups with over a hundred cows such recognition is not possible and more bullying occurs. The behavioural problems in moving cows from one group to another are discussed in Chapter 5.

Herd Movement

Cattle moving of their own volition tend to string out in a line with natural leaders at the front. When a herd is driven, for example along a trackway, the ones at the back are under some pressure so that the herd forms more of a triangular shape. Stress is caused if cattle are rushed at a pace faster than the normal level of the slower members of the group. It is easy to damage feet and udders and cause poor let-down if the cows are rushed just before milking. A concrete road has obvious advantages but if a build-up of dung and soil is allowed, especially if the soil is stony, then increased incidence of lameness can be expected.

Ideally, a milking herd comes to the gate when called, avoiding the need to drive them from a field or yard or to use a dog for collection. Few dogs are sufficiently gentle when working with cattle, and have a tendency to nip heels and cause too much disturbance. Gathering cattle from a large field can be time-consuming and frustrating but the author is not impressed

Moving towards afternoon milking. Rushing cattle along tracks can damage feet and udders and create stress.

by herdsmen who dash around the field on a tractor trying to out-perform a fast dog!

Separating one animal out of a herd, such as a down-calver from a group of dry cows, is another operation requiring considerable skill on the part of the herdsman. A stick is a useful aid on such occasions, not for hitting the animals but to show them that you do mean business, and it should always be kept down because a stick held high seems to alarm the stock. If the animal you are trying to separate is nervous it may be necessary to take more than one out of the group so as to accompany her to the new location.

An inexperienced handler spends too much time running after cattle, trying to recover a situation from which they have escaped. The skill comes in avoiding a breakaway by watching the group carefully, working and moving quickly but without too much noise. Cattle move easier if they have only one obvious way to go. Loading into a cattle truck, for example, is made much simpler if the vehicle can be reversed up to a race which has high sides so that there is no chance of the animals jumping over. A little straw on the ramp usually helps reduce the fear of loading, as does allowing them time to smell their way into the 'unknown'.

Teaching to Lead

Show cattle and young bulls need to be halter trained and this is best started when they are young calves. At this age they can easily be tied up with light halters to a ring in a wall, so that they become used to being restrained.

Once accustomed to a halter, they continue to respond to safe handling in this way as adult cattle. When fitting a halter it is best to put the poll piece over the ears first, before tightening the loop around the nose. Bulls of course need to be handled from yearlings with a pole as described in Chapter 11.

Casting

This method of handling cattle is now rarely used. Even when operations have to be carried out on the feet a modern crush is preferable for restraining the animal. However, a good stockman should know how to cast cattle using the Reuff method. The animal is first haltered so that the head can be controlled, preferably by an experienced person. A long rope is used, first

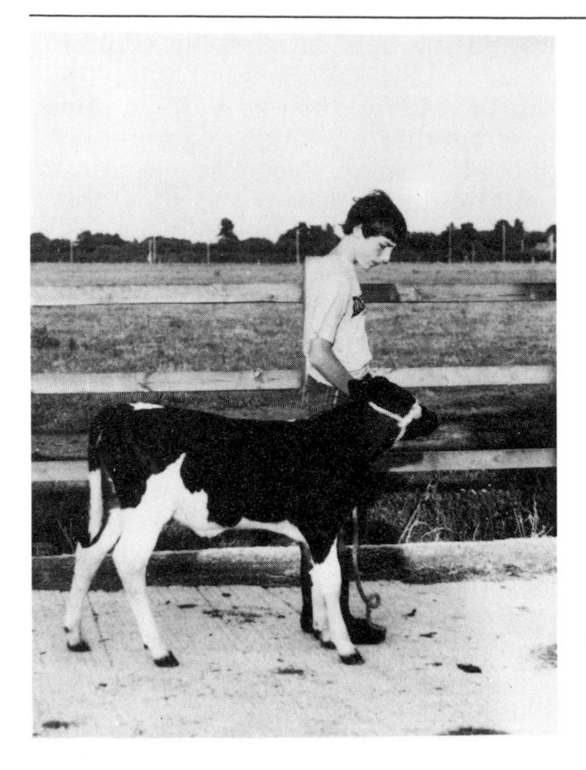

Halter training a young heifer. Show cattle and young bulls need to be halter trained, and this is best started at the calf stage.

placing a loop round the neck and securing it above the withers. The long end of the rope is then taken along the back forming a half-hitch around the chest just behind the front legs and another around the abdomen in front of the udder in a cow. If one or two

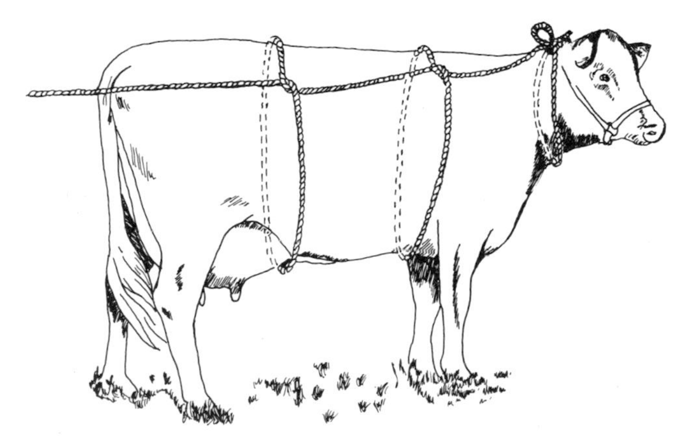

Figure 1. Reuff's method of casting cattle.

people then pull steadily on the rope the animal should collapse to the floor. The knot used in Reuff's casting is a quick-release bowline.

Using the Crush

This is the most satisfactory method of restraining cattle. Metal or wooden stocks hold the head firmly and a bar is placed behind the rump. Most animals tend to be nervous when they enter the crush so that quiet but firm handling is required. It is a good idea to let the animal entering see the one that is leaving. The approach race should be strong, safe and free from any protrusions, which could injure the cattle. The floor of the race, especially if concreted, should be kept clean so as to minimise slipping. The floor of the crush is often constructed of timber, especially if the equipment doubles-up as a weighbridge; this certainly needs to be kept clean and in a good state of repair. Moving parts like the gate hinges and any weighing mechanism should be oiled frequently.

It is essential when using the crush to be alert and avoid being trapped, especially when releasing an animal from the front gate.

Specialised crushes for foot trimming work will be described further under the topic of lameness in Chapter 8.

Restraining Without a Crush

It is often necessary to restrain a beast in a location where there is no access to a crush. If an assistant is available it may be possible to use a lightweight gate to trap the animal against a pen wall; then with thumb and first finger take a firm grip of the nostrils. Fingernails obviously need to be short to avoid damaging the tissue of the nostril, and if the animal is mature, considerable strength is required!

Drenching

This is a job best carried out using a crush, although the nostril grip will be adequate with the occasional animal. The medicine to be administered should be placed in a long-necked plastic bottle. If you are right-handed, you should stand on the right-hand side of the animal, restraining its head with the left hand. The bottle is

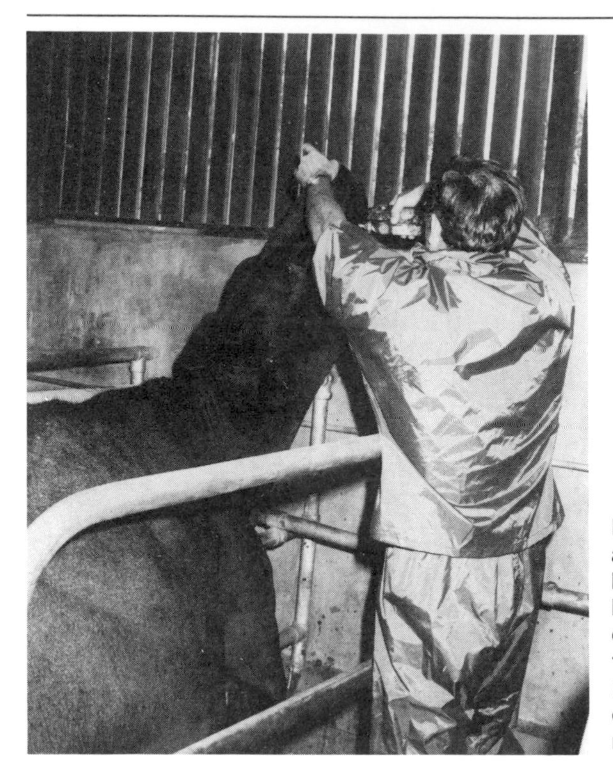

Drenching a cow. If a large group has to be dosed, it is a job best carried out in the crush, but when only the occasional animal requires treatment, catching with the nostrils is adequate.

placed in the side of the mouth above the tongue and just behind the incisor teeth. The fingers of the left hand are placed in the same gap at the opposite side so encouraging the mouth to open. Care has to be taken to allow swallowing, giving a pause between each dose.

A range of special drenching guns are now available. Some have a crooked nozzle which is ideal for delivering the drench well back over the tongue, without even having to restrain the head.

Passing a Stomach Tube

When dealing with bloat in a calf (see Chapter 8), it may be necessary to release gas from the stomach by passing a length of flexible garden hosepipe by mouth into the oesophagus. This needs to be passed with care over the top of the larynx and trachea. To check that it has not been passed into the trachea, listen to the end of the pipe. If air is moving in and out in tune with breathing, then the tube needs to be removed and re-inserted.

Injections

This is a method being increasingly used to administer drugs and one in which the herdsman needs to be particularly skilled. Care and attention to cleanliness, knowledge of the correct type and site of injections and selection of appropriate needle are all essential. Initial supervision by a veterinary surgeon or training officer is advisable before undertaking the task alone.

The subcutaneous (under the skin) method is commonly used and in cattle is ideally carried out at the base of the neck, in front of the shoulder. The site should first be swabbed with skin disinfectant, a pinch of skin taken and the needle carefully inserted. After injecting the medicine and withdrawing the needle a gentle massage of the area will help disperse the material from the site.

Intramuscular (into the muscle) injections are usually carried out into the buttocks. The needle should be detached from the syringe and held firmly in the finger and thumb. A few firm taps on to the site (previously disinfected) with the back of the fist prepares the animal so that when the needle is plunged into the muscle it is not frightened.

Intravenous (into a vein) injections do need rather more skill and training. The jugular vein along the side of the neck is a good site

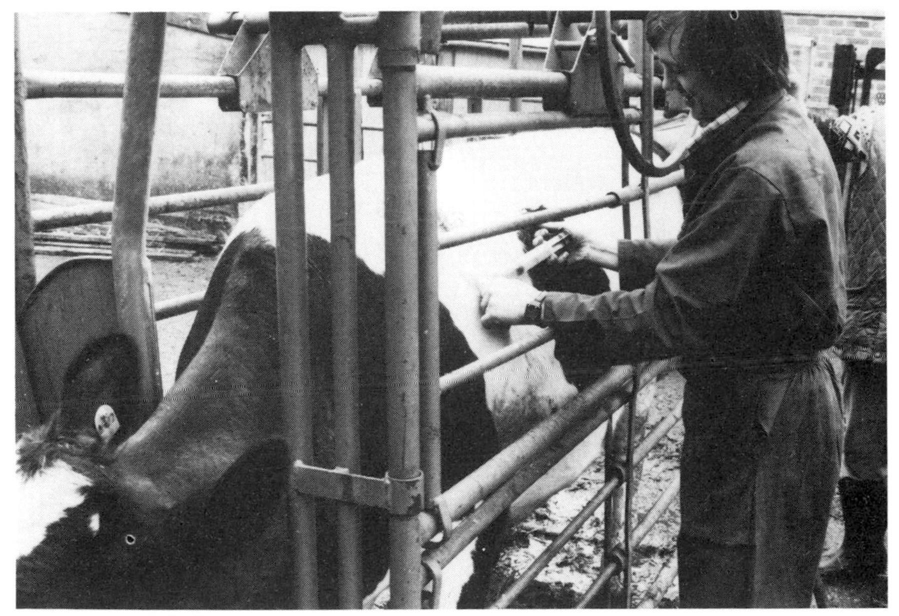

A subcutaneous injection for the control of foul-in-the-foot.

to select. A point along the neck should be clipped and swabbed with spirit. With mature cows and bulls it is necessary to use a rope noose which, when tightened, distends the vein. A short, sharp, bevelled needle can then be inserted into the vein, syringe attached and medicine pushed steadily into the bloodstream while the rope is released. After you remove the needle hold the thumb firmly over the site to prevent any loss of blood.

Inter-uterine (into the uterus) injections are frequently required to control metritis. The insertion of a catheter is usually a job for the vet; when undertaken by herdsmen it needs to be done with much care to avoid tissue damage.

DAILY ROUTINE

An example of a herdsman's daily routine, during the winter months in a one-man unit involving parlour milking, cubicle housing and self-feeding of silage.

Task

1. On arrival, check the down-calving cows. If an animal needs assistance allow 5–10 minutes to deliver the calf; otherwise call for help to avoid undue delay in milking (especially if the milk tanker comes early).
2. Final preparation of the milking equipment including check on bulk tank temperature (main preparation previous evening).
3. Bring cows to collecting yard checking on any cows in oestrus.
4. Switch on the vacuum pump, check the gauge and undertake milking; this will usually include concentrate feeding.
5. Wash parlour and sterilise milking equipment.
6. Check breeding/health records of cows diverted for treatment. Phone AI centre to order inseminations.

Breakfast

7. Scrape slurry from cubicle passageways. Check water troughs. Top-up bedding in cubicles (alternate mornings).
8. Feed and bed-up dry cows.
9. Provide a mid-morning manger feed to milking cows (sugar beet pulp, concentrates). Observe cows while they are eating,

then scrape slurry from self-feed area. Adjust feed barrier or fence.

10. Assist vet or AI inseminator. Treat other problem cows, e.g. lame ones.
11. Feed calves on milk substitute (using once-a-day system).
12. Late morning check for cows in oestrus (10–15 minutes).

Lunch

13. Check once again the down-calvers.
14. Complete the washing/disinfection of the bulk milk tank following collection and activation of automatic washer by tanker driver. Prepare parlour for milking, replacing any damaged rubbers.
15. Undertake any essential maintenance such as a broken cubicle rail.
16. Complete the records for the day, i.e. AI, calvings, vet, mastitis. Ear-tag calves and sketch pedigree cards.
17. Bring cows to collecting yard then scrape slurry from cubicle passageways.
18. Afternoon milking, concentrate feeding.
19. Check down-calvers, dry cows and calvers.

Tea/evening meal

20. Late-night check of all stock, especially oestrus detection (20–30 mins.)

Sleep well!

ANNUAL ROUTINE

An example of the routine tasks undertaken by a herdsman during the year, from an early autumn-calving herd with the production year beginning in June.

JUNE Provide data for completion of MAFF June 4th returns.
Start drying-off early season calves.
Carry out essential maintenance of buildings and equipment.
Visit other farms to inspect progeny of bulls which could be used in the herd.

Brief relief herdsman in preparation for forthcoming holiday.

JULY Assist as necessary with 2nd cut silage campaign.
Summer holiday—suitably refreshed for the new production year.
Dose yearling heifers against internal parasites.
Order semen for next breeding season.

AUG. Start calving season for heifers (22–24 months old).
Receive annual straw requirements and arrange for a quantity to be chemically treated for feeding purposes.
Assist with reseeding of several grass paddocks.
3rd cut silage—into big bales and wrapped.

SEPT. Meeting with feed advisers to plan 'winter' rations.
Start of main calving season for cows.
Cull the cows that failed to conceive and have gained weight at grass.
Visit European Dairy Farming Event at Stoneleigh to keep abreast of new technology.

OCT. Meeting with AI centre manager to finalise semen requirement for forthcoming breeding season.
Open a silage clamp and slowly introduce 'winter' diets to freshly calved group.
Attend local Herdsman's Club meeting on 'Breeding the Cow for the Future'.

NOV. Meeting with vet to finalise plans for his input during breeding season.
All cows and youngstock housed for winter months.
Arrange for prostaglandin treatment of heifers and double AI.
Start of the breeding season for cows.

DEC. Meeting with herd owner to discuss possible replacement of equipment for next silage season (Smithfield Show).
Make preparations to minimise daily routines over Christmas period.
Apply tail paint to all cows not seen in oestrus by 40 days of lactation.

JAN. Short break over New Year (most calvings completed).
Check stand-by generator in case of power-cuts due to adverse weather.
Meeting with herd owner and costings adviser to discuss annual performance to date.
Implement culling and feeding plans to meet Quota production levels.

FEB. Plan grassland use for forthcoming grazing and conservation season.
Check lime status of cow paddocks.
Apply 1st nitrogen application as weather conditions allow.
Continue pregnancy testing with the vet.
Check that order is placed for silage additive and plastic sheeting.

MAR. Arrange for slurry store to be emptied.
Check field fences and water troughs.
Cull further cows when prices are appropriate (especially if winter feed stocks are low).
Arrange for fields due to cut for silage to be flat rolled.

APR. Turn out heifers, then cows by day.
Slowly adjust diets as grass growth improves.
Vaccinate heifer calves for husk.
Cows out by night by end of month.
Commence 'spring cleaning' of buildings.
Spring break before silage making.

MAY Prepare silage clamps for new season's crop.
Pressure wash and disinfect cubicle beds.
Continue to apply fertiliser after each paddock grazed.
Assist with 1st cut silage.
Turn out calves to 'clean' ley.
Fit fly-control ear tags to in-calf heifers.

Much satisfaction can be obtained by undertaking these husbandry tasks in a correct and safe way. Care of all the equipment is essential. When each task is completed, tools and equipment should be cleaned and returned to the store and injection needles should be sterilised ready for use on the next occasion.

The reader of this chapter may now appreciate why its author claimed at the outset that there is *some* truth in the saying that good stockmen are born, not trained—but that, equally, some important qualities and skills can be gained by experience and training. More often than not, good stockmanship exists precisely because of that kind of combination of natural and acquired skills, and the aim of this chapter has been to demonstrate that fact.

CHAPTER 2

PRINCIPLES OF FEEDING

Another major factor affecting the profitability of a cattle enterprise is the provision of feed for the stock which is adequate in both nutritional and economic terms. Carefully controlled feeding experiments have been undertaken to obtain the precise nutritional requirements of all classes of cattle. Costings schemes have been worked out including computer programmes which, when provided with information regarding available feeds and prices, calculate optimum or least-cost rations.

Despite all these developments, there continues to be a most important role for herdsmen in modifying such 'standard' figures to meet the day-by-day or even hour-by-hour needs of his stock. An animal which has been 'off-colour' may need to receive a special diet for several days before full recovery takes place. In order for the herdsman to undertake the feeding of the herd effectively and with confidence, it is necessary for him to first understand some of the principles or science of feeding.

The Digestive System

Cattle use their food for two main purposes: first to provide the means of maintaining life and normal activity; this is known as the maintenance requirements. Secondly, for the production of milk, meat and the development of the unborn calf. Following digestion, the food provides the body tissues with energy, protein, minerals, vitamins and some water (especially from grass, kale and silage). It will therefore be useful, at this juncture, to consider briefly how the digestive system does in fact work.

In essence it consists of a long feed tube passing from the mouth to the anus as shown in Figure 2. The food is broken down in a series of processes: firstly mechanical action, primarily chewing; secondly microbial fermentation, particularly in the rumen; and, thirdly, biochemical digestion, involving enzymes, which takes place at various parts of the gut. The nutrients or products of digestion pass through the wall of the digestive tract into the blood stream for transport to other organs such as the udder for the production of milk.

Food is taken into the mouth by a grasping and tearing action using the lips, teeth and tongue. The rear molar teeth then grind the food before swallowing, as they do again when a bolus of regurgitated food is further chewed during rumination (cudding). From the mouth, food passes down the oesophagus to the first of the four stomachs known as the rumen. It is the largest of the four and is attached to the much smaller reticulum which is located in front of the rumen. Also opening from the rumen and lying to the right is the small round omasum which is followed by the abomasum or true stomach.

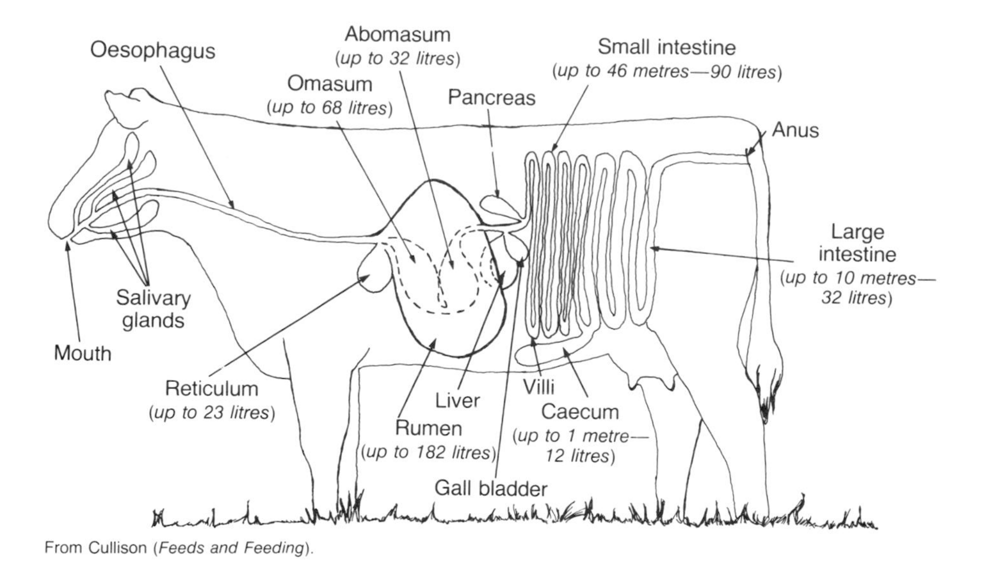

From Cullison (*Feeds and Feeding*).

Figure 2. The digestive tract of the cow.

The *reticulum* and *rumen* work together as the main fermentation site. The vessel contains millions of microbes which are involved in breaking down the food into a fine liquid mass. Some of the larger pieces of food are returned to the mouth during rumination by reverse action of the muscles. Fermentation is a continual process, hence the need for regular feeding to keep the microbes working to the full. Large amounts of saliva are required in fermentation, and this is produced from the glands in the mouth (up to 150 litres per day in a high-yielding cow). This provides water to assist the chewing process and salts to buffer the acids formed from the breakdown of carbohydrates. Gases such as carbon dioxide and methane are produced during fermentation and in the healthy animal are removed by belching or during rumination. Bloat becomes a problem in an animal which is unable to remove gases in this way. A regular supply of fibrous food keeps the vessel in good function and helps to prevent bloat.

Most of the carbohydrates (energy-supplying substances) are broken down in the reticulo-rumen to volatile fatty acids such as acetic, propionic and butyric acid. Simple sugars are broken down rapidly, whereas cellulose and hemi-cellulose from the cell walls of plants are degraded much more slowly.

Modification of oils and fats also takes place in the reticulo-rumen and some products of fermentation are absorbed into the blood stream, but most 'feed' the microbes which then pass down the alimentary tract for subsequent digestion.

The *omasum* has a lining of tissue similar to the pages of a book. Its main function is to remove surplus water from the material before it flows into the abomasum where the main biochemical digestion begins.

The *abomasum* secretes large amounts of acidic gastric juices which kill the microbes and start the process of protein digestion. The material then flows into the U-shaped duodenum where it receives enzymes from the pancreas and bile from the gall bladder to neutralise the acidity of the residual gastric juices.

The *small intestine* is extremely long (up to 46 metres). It is the main area of protein absorption and the site for some glucose production from the digestion of carbohydrates.

The *colon* or *large intestine* is not as long as the small intestine but has a bigger diameter. Excess water is removed from the gut contents at this point, so concentrating the faeces. Some micro-organisms are at work in the colon, further fermenting fibre particularly in the caecum which acts as a secondary fermentation chamber.

Another major organ involved in food digestion is the *liver*. It acts as a store of energy, deals with ammonia in the blood and converts numerous complicated chemicals into much simpler ones for use by other body organs. The *kidneys* too have an important role in taking waste products from the blood, these being discharged in the form of urine.

It can be seen that, apart from eating, defecating and urinating, which take place at intervals, digestion is a continuous process. Any part that suffers from an upset creates problems to the whole

Yearling Hereford × Friesian steers eating maize silage at a feed fence. The diagonal rails of the fence deter the dominant animals from backing out to change feeding positions when the concentrate part of the ration is fed.

system. Herdsmen therefore have an important role in providing feed, water and the right environment so that efficient digestion takes place.

Feed Requirements

Since 1975 the method of expressing feed requirements in the United Kingdom has been based on the Ministry of Agriculture Technical Bulletin No. 33, which replaced the previous starch equivalent system. The system uses the metabolisable energy (ME) in the food as a basis for formulating rations. The ME is the energy remaining after the loss in faeces, urine, gases and body heat. The basic unit used to measure the energy is the megajoule (MJ).

The basis of a rationing system is to meet the energy needs of the animal from the foods provided, but also taking into account the digestible crude protein (DCP) requirements. The requirements of ME and DCP for maintenance are supplied in Table 2.1 and for production of milk in Table 2.2. The ME and DCP considerations in liveweight gain and loss are supplied in Table 2.3.

Dry Matter Intake

The next consideration is an important one, being the amount of food that the animal will eat. Appetite has been shown to vary with the size of the animal, its level of production and also the digestibility of the ration, its palatability and the manner in which the food is presented. A guide to dry matter intake (DMI) of cows at different levels of yield is supplied in Table 2.4.

Table 2.1 Maintenance only

L.wt. (kg)	*ME (MJ/day)*	*DCP (g/day)*
300	36	200
350	40	250
400	45	250
450	49	300
500	54	300
550	59	350
600	63	350
650	68	400

Table 2.2 Production of 1 kg milk

Type of milk	*B.F. %*	*Protein %*	*ME (MJ/day)*	*DCP (g/day)*
Friesian	3.6	3.2	4.98	60
Ayrshire	3.8	3.3	5.17	65
Channel Island	4.8	3.7	5.93	70

Table 2.3 Live weight change

Gain/loss	*MJ (kg/day)*	*DCP (g/day)*
+1 kg	34	320
−1 kg	28	—

Table 2.4 Probable dry matter intake in early lactation (weeks 1–10)

		Milk Yield (kg/day)				
L.wt. (kg)	*Dry*	*10*	*20*	*30*	*40*	*50*
350	4.8	8.3	11.8	15.3	18.8	22.3
400	5.4	8.9	12.4	15.9	19.4	22.9
450	6.0	9.5	13.0	16.5	20.1	23.9
500	6.6	10.2	13.7	17.2	20.7	24.2
550	7.3	10.8	14.3	17.8	21.3	24.8
600	7.9	11.4	14.9	18.4	21.9	25.4
650	8.5	12.0	15.5	19.0	22.6	26.1

Intake is actually dependent on the speed at which food passes through the digestive tract. Low-fibre, laxative types of diet such as spring grass pass through so quickly that digestion is inefficient and the animal may lack sufficient nutrients. High-fibre diets, on the other hand, can be so indigestible that again the requirements of energy and protein fail to be met.

Calculating a Ration

In this example a cow of 550 kg liveweight is producing 30 kg milk at 3.6 % BF and 3.2 % protein. She is in early lactation and is losing weight at 0.5 kg per day. (See composition of feeds, Table 2.5.)

Requirements

	ME (MJ)	*DCP (g/day)*
Maintenance (550 kg)	59	350
Milk production (30 kg)	149	1800
Liveweight loss (0.5 kg)	(14)	0
Total	194	2150
Probable Dry Matter Intake	17.8 kg	

Ration

	Fresh wt. (kg)	*DM (kg)*	*ME (MJ)*	*DCP (g)*
Grass silage (high digestibility)	40	10	103	1050
Wet sugar beet pulp	5	0.9	11	60
Concentrates (standard)	8	6.9	86	1104
	53	17.8	200	2214

This diet adequately meets the energy and DCP requirements of the cow and achieves this within the probable level of dry matter intake.

Protein Degradability

Although, as explained above, the DCP method of calculating the protein requirement of cattle is widely used, a new system has been introduced which takes into account the degradability of the protein in the ration. It is a superior system of calculating the requirement, especially for high-yielding cows which have been shown to benefit from protein which escapes degradation in the rumen and is absorbed as amino-acids in the hind gut. These types of protein are known as undegradable (UDP), good sources being fish meal, meat and bone meal and grass silage which has been treated with formaldehyde. Foods with medium degradability are dried grass, soya bean meal and maize silage. Those which are degraded in the rumen (RDP) are used by the micro-organisms, common sources being grass, hay and silage without additive. The rumen micro-organisms can also utilise non-protein nitrogen (NPN) such as urea which has for some time been fed to beef cattle but is now being incorporated into dairy diets.

Table 2.5 The composition of some common cattle feeds

Feed	*Dry Matter (g/kg)*	*ME (MJ/kg DM)*	*DCP (g/kg DM)*
Grazing (rotational)	200	12.1	185
Grazing (extensive)	200	10.0	124
Silage (high D)	250	10.3	105
Silage (moderate D)	250	9.8	102
Hay (high D)	850	9.0	58
Hay (moderate D)	850	8.4	39
Straw (spring barley)	860	7.3	9
Straw (winter barley)	860	5.8	8
Kale	140	11.2	106
Mangels	120	12.4	54
Potatoes	210	12.5	47
Barley	860	12.9	82
Oats	860	12.0	84
Maize	860	14.2	105
Sugar beet pulp (wet)	180	12.7	66
Brewer's grains (wet)	220	10.0	149
High energy dairy compound	860	14.0	180
Standard dairy compound	860	12.5	160

If sufficient RDP is not available, the rate of fermentation of fibrous as well as concentrate rich diets will be reduced. This leads to a reduction in intake, lower energy supply and reduced milk production.

It is therefore important to have the optimum balance of UDP and RDP in the diet. Figure 3 illustrates the increasing requirements for both as milk yield increases.

To obtain the appropriate levels it is necessary to calculate

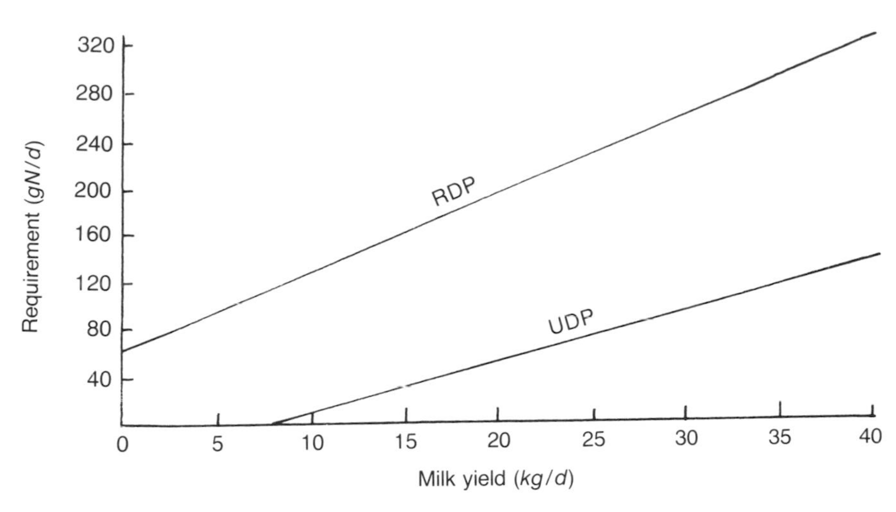

From: Wilson and Brigstocke

Figure 3. Protein requirements related to milk yield.

the total energy and energy concentration in the diet for the desired level of yield. Then determine how much total protein is required to match that energy. After calculating how much can be provided by the microbes, the remainder has to be provided in an undegradable form. The feed trade and other advisory organisations have computer programmes which deal with this aspect of ration formulation.

Mineral Requirements

The major mineral requirements for dairy cows are calcium and phosphorus. The calcium:phosphorus ratio is important, as an imbalance can cause infertility. The requirements for maintenance and production of a 6,000 litres lactation are 50 g calcium and 70 g phosphorus per day. There are considerable reserves of both elements in the skeleton. However winter diets can be short of calcium and pasture deficient in phosphorus. Feeds such as kale and sugarbeet are high in calcium and low in phosphorus (10:1). The aim should be a ratio between 1:1 and 2.5:1.

Shortages of sodium are rare but salt licks are useful, especially when high rates of potassium fertiliser are used. Magnesium is frequently required, at the rate of 50 g per day, to prevent hypomagnesaemia.

Vitamin Requirements

The two vitamins likely to be in short supply are A and D in the winter months. Vitamin A is stored in body tissue from summer grazing; kale and high-quality silage are also good sources. Vitamin D is provided in well-cured hay, and concentrates are often supplemented with synthetic A and D.

Vitamin E and/or selenium deficiency due to feeding poor hay to calves can cause muscular dystrophy. When the animals are turned out to grass, they suddenly take more exercise and can develop a stiff gait. Veterinary advice on supplementation should be obtained, as selenium is very toxic.

Before concluding the principles of feeding and moving in the next two chapters to the practical aspects of feeding in winter and in summer, it is necessary to stress the long-term effects of feeding a dairy cow. The feeding level at one point of the lactation can have a marked effect upon yield and body condition, not only later in that lactation but also in subsequent lactations. This is well illustrated by good feeding in late lactation which increases body condition at calving so that a higher peak yield is achieved as well as a higher total lactation yield.

Figure 4 demonstrates the relationship between milk yield, body weight change and appetite. The shape of the lactation curve is affected by genetic potential and also by feeding and

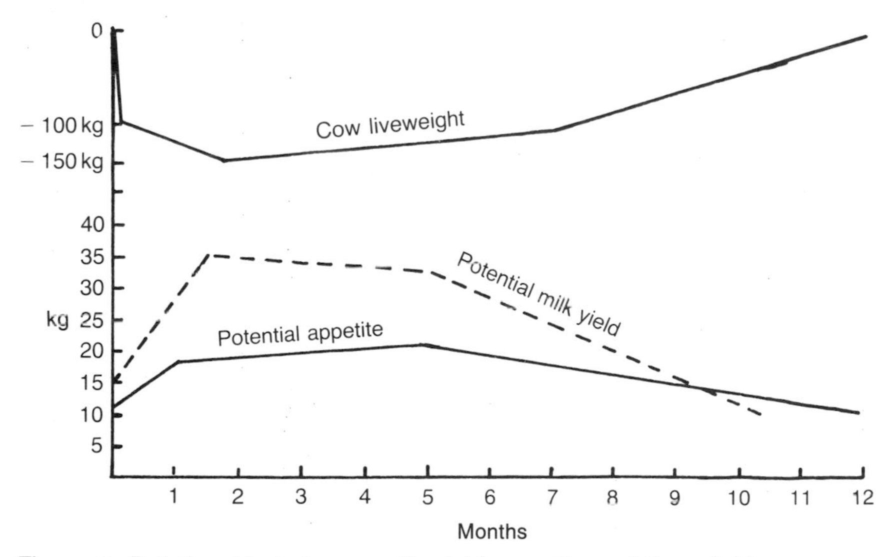

Figure 4. Relationship between milk yield, appetite and liveweight.

management. The peak yield is critical, because for each litre obtained at peak the overall lactation increases by 200 litres. Rate of decline after peak is normally 2.5% per week so that the yield around the 100th day of lactation is often a better indicator of total yield than is peak yield.

Appetite is lowest at calving and it increases to a peak around the fourth month of lactation before declining as the developing foetus takes up body space. The energy deficit from the feed in early lactation can be compensated by the mobilisation of body fat in the first two months after calving. It is therefore necessary to ensure that cows are in good body condition at calving. The regaining of weight before the subsequent calving is more efficiently undertaken while the cow is still milking rather than when dry. Overfeeding during the dry period can also lead to dystokia.

In conclusion here, the importance of correct feeding as a major determinant of profit, cannot be overstressed. Recent figures derived from the regularly conducted National Investigation on the Economics of Milk Production indicates that total food

Milking cows retained in self-locking yokes to allow individual rationing of concentrates according to yield. Note the large plastic ear-tag used for identification.

costs—grazing, conserved bulk foods and concentrates—account together for some 56% of all costs of producing milk. The significance of a figure of that magnitude—and therefore of this chapter—is clearly self-evident.

CHAPTER 3

WINTER FEEDING

The last chapter dealt with the theory of feeding and how diets are calculated on paper. Now we move to the most relevant topic for herdsmen, that of feeding in practice, starting first with the winter period.

Before describing alternative systems in use for the feeding of forages and concentrates, it is necessary to outline several points of general stockmanship involved in the stockmanship of feeding.

- Arrange regular feeding times (including weekends and when relief staff are on duty). Remember that cattle are creatures of habit and that the digestive system works more effectively if a regular pattern of feeding is maintained. As regular rumination is also required, rest periods between feeds need to be provided.
- Avoid over-feeding at one meal, especially with concentrates to high-yielding animals. The ideal 'meal' is first concentrates, then hay or a fibrous feed, followed by succulents.
- Allow the appetite of the animals to be satisfied. The theoretical diet may well provide essential nutrients but fail to fill the cattle, so that they remain restless and perform below expectation. Liberal straw bedding helps contentment as it allows hungry animals to eat some of the material before it becomes fouled.
- Allow time to observe the animals while they are feeding. Check that adequate trough space is available (recommendations given below), that minimum bullying is taking place and, unless an ad-lib system is in use, that every animal comes to feed.
- Check that the expected production level is being achieved. This involves measuring the production by recording the yield of milk or the weighing of growing stock. Recording of individual milk yield is ideal but much can be gained from

Milking cows eating high-quality silage from a manger. This specially designed barrier is comfortable for the cattle to use as it slopes out from vertical, so allowing improved reach of the feed.

Cleaning out waste material from a concrete manger. It is necessary to ensure that diets are palatable and so, for example, when low-quality silage has to be offered, refusals need to be removed regularly. Note the highly effective freeze-brand mark of 732.

using group yields (where applicable) or even daily herd yield. Adjustment of the diet will be required to correct variations from target. Note that feeding levels also affect the quality of production, e.g. butterfat and protein of milk.

- Ensure that diets are palatable. Food refusals may not be due to low appetite but to such problems as mouldy hay, dirty roots or butyric silage. Take trouble to prevent problem material, e.g. waste from the top of a silage clamp, being loaded into forage boxes or being allowed to drop down to a self-feed face. Clean out mangers regularly if low-quality material has to be offered. Small quantities of highly palatable materials such as molasses can be used to improve the palatability of a diet.
- Check the level of digestibility of the diet by the consistency of the dung, avoiding undue scouring or constipation. Buffer lush grass or kale with a little hay or clean straw.

Loading a complete-diet feeder. The feeder is fitted with electronic weighing equipment so that each constituent of the ration can be accurately added.

- Watch out for laminitis and sore feet in high-yielding animals which are receiving a concentrate-rich diet. The addition of sodium bicarbonate to the diet (250–300 g per day) has been shown to be beneficial and it can also help a low butterfat problem.

- Avoid abrupt changes to the diet. The pattern of fermentation in the rumen is seriously affected especially if the concentrate: forage ratios are abruptly changed. Arrange for a change of diet to take place over several days.
- Minimise feed wastage, especially of expensive concentrates, but also of hay, silage or complete diet. Avoid over-filling low-capacity mangers, adjust the height of mangers where applicable and always drive accurately when dispensing from a forage box—even on Sunday mornings!

Planning the Feeding Programme

It is advisable to plan the feeding programme well in advance of the winter period, by first evaluating the available stocks of home-grown feeds. Any requirement for purchased feeds to supplement a shortfall can then be arranged in good time, hopefully avoiding the need to buy at the end of the period when prices are usually at the highest level.

Once the home-grown 'background' feeds are stored in a clamp, tower or stack, or as a standing fodder crop in the field, there is little the herdsman can do, at this stage, to influence feed quality. Ideally, he will have used his influence and helped out earlier in the season to obtain stocks of the optimum quality, by for example, rolling mole hills to avoid soil contamination of silage and by ensuring thorough sealing of the clamps.

The first part of the planning exercise is to quantify the amount of each type of feed available. With baled hay and straw, a count may need to be taken unless accurate records have been kept by the baler operator or by the staff involved in stacking. It is then necessary to weigh a representative sample of the bales and, by multiplication, to obtain a total weight. Few difficulties should arise weighing traditional small bales but it may be necessary to travel to a weighbridge to obtain the average weight of a few large bales.

Evaluating silage quantities is rather more difficult due to the fact that uneven shaped clamps are frequently formed. It is necessary to obtain the volume of material (m^3) by using a measuring tape (metres) and multiplying length $\times$ width $\times$ average depth.

The average weight of one cubic metre is 660 kg, although shortchopped material in deep clamps is heavier, and unchopped material in shallow clamps is lighter.

Feed Requirement

The length of the average winter feeding period for budgeting purposes is usually 180 days although in northern regions it may well be over 200 days. Assuming that the home-grown background feeds are of sufficient quality to provide the nutrients for maintenance plus 5 litres (M + 5), then the appropriate requirements are:

Cow live wt. (kg)	*Silage (kg)*	*Hay (kg)*
600	7,650	2,550
400	6,000	2,000

More precise quantities will depend upon such factors as herd yield and attitude to concentrate feeding. High-yielding cows require less forage but animals offered only low levels of concentrate, but at the same time having easy access to high-quality silage, will eat as much as 8–9 tonnes each during the winter months.

If alternative foods are available a useful conversion is:

1 kg hay = 3 kg silage
= 4 kg kale or beet tops
= 5 kg mangolds
= 2 kg straw
= 1.5 kg chemically treated straw
= 0.5 kg straw + 0.5 kg straw balancer
= 1 kg brewer's grains + 0.25 kg cereals

Forage quality is best obtained from laboratory analysis, average value being given in Chapter 2, Table 2.5, although a useful guide to good material is:

Grass silage Greenish-yellow colour
leafy, few flowering heads (cut early)
sharp acidic smell
not over-heated or with a butyric smell
well made, absence of soil contamination
difficult to squeeze out moisture by hand

Maize silage	material with a good grain content
	dry matter between 27 and 30%
	short uniform length of chop
	high proportion of the grain shattered
	absence of large pieces of unchopped leaf
	not warm (which would indicate spoilage)
Hay	absence of mould, a 'good' smell
	green or yellowish-green colour
	little loss of leaf
	no seed formed (cut early)

Once the amount and quality of the available feeds are known, rations can be formulated for each class of stock, as explained in the last chapter. It is worthwhile having a computer calculation undertaken by a feed company or by an organisation such as ADAS or Genus. This is particularly useful if 'other' feeds like beet pulp, brewer's grains or arable by-products are to be purchased. Given the prices of such feeds, the programme will calculate which one, or even a combination, is the 'best buy'.

FORAGE FEEDING SYSTEMS

1. Silage

There are a number of alternative methods of feeding silage from store, the suitability for a particular situation depending upon such factors as building layout, herd size, yield level and attitude to mechanisation of the management and dairy staff.

Self-feeding This is a satisfactory method of feeding silage, especially for medium-sized herds using minimal labour, providing that the layout is well planned and the slurry disposal system is satisfactory.

Short-chop material is ideal; it should not be stacked more than 2 metres deep unless some hand-cutting and forking down of the upper layers take place. The optimum allowance of feeding face depends upon cow size but primarily upon the time allowed for access. An adequate allowance for a small cow with 24-hour access is 20 cm but 80 cm will be required if large animals have to be batch fed at the face.

Herdsmen usually control the face by using a good 'hot'

electric fence or feed barrier to restrict the cow's access. Some research has indicated that an increase in silage intake is obtained when a solid bar is used as a barrier rather than an electric fence. The barrier or fence should be moved forward at regular intervals based on knowledge of the performance of the stock and the availability of supplies. It is helpful to use marks along the walls of the clamp as an aid to monitoring use. When quality is low, it may pay to allow some rejection and remove the uneaten material daily for feeding to less critical animals. Problems can arise if two or more silage cuts have to be ensiled in the same clamp, particularly if the layout does not allow the first cut to be utilised when required for freshly calved animals.

In order to prevent silage effluent flowing from the clamp while and after the material is ensiled (very high pollution potential), some farmers use a layer of absorbent material such as chopped straw or sugar beet pulp at the base of the clamp. There is considerable skill in judging the optimum depth of the layer, particularly for self-feeding. Too much adds to costs and dry straw is not particularly palatable whereas too little allows leakage—of an even more concentrated effluent. In situations where effluent can be safely collected and stored, it is possible to feed the liquid to stock, ideally in restricted quantities.

Heifers in their first lactation which are in the process of losing their calf teeth need to be watched carefully, especially if they have to pull out unchopped silage. Extra concentrate supplementation will be required for such animals until their broad teeth are fully cut.

Increased flexibility in cow management is possible with self-feeding if more than one feeding face can be provided, so allowing the herd to be subdivided. Such a facility enables the newly calved and first-lactation heifers to be managed in one group; it reduces stress on these critical animals, enabling higher peak yields to be obtained, as well as better conception rates.

Self-feeding is only suitable for high-yielding herds if facilities are available to feed concentrates other than in the milking parlour, i.e. in a manger or out-of-parlour feeders.

Mechanical handling With large herds, or on farms where the staff have considerable mechanical interests and skills, the handling of silage out of clamps by machine is a practical alternative. Many farms use a standard front-end tractor loader with a normal fork or bucket, others have a special grapple fitted

Self-feeding of silage.

A materials handler taking maize silage from the clamp. If secondary fermentation or rotting is to be avoided, this must be done with minimum disturbance to the remaining silage.

to the loader, but some use specialised silage block cutters. With all types of machine the aim should be to remove the required amount of material with minimal disturbance to the remaining face. Secondary fermentation (or rotting) soon takes place when air is allowed into the face. The problem can be reduced by working methodically across the face of the clamp from one side to the other, by re-sheeting daily and by spraying the disturbed face with propionic acid from a hand-spraying unit.

All silage handling machines, but particularly block cutters, need to be regularly checked and serviced, as reliability is an important feature of the system. A number of larger farms use industrial type loaders with the objective of increasing reliability as well as output.

When loading into a forage box or a complete diet feeder it is vital to avoid damaging the unloading and mixing mechanisms by first teasing-out the blocks and large forkfuls of silage.

Big bales The development of big-bale silage is proving of considerable benefit to many beef enterprises but also to some dairy farms, especially smaller ones which have difficulty in justifying the capital cost of traditional silage making. High-quality material can be obtained by cutting and baling early, usually using a contractor, and by giving care and attention to sealing and stacking the bags, and to preventing vermin chewing the plastic sacks.

The availability of a quantity of bagged silage is proving, on many farms, to be a convenient method of buffer feeding cows during the summer and autumn months. Instead of opening a clamp to supplement inadequate grazing, large bales of silage can be fed from mangers located on concrete in the yards or even at pasture.

2. Hay

Hay is a most convenient form of conserved forage to feed, particularly in cowsheds but also in loose-housed situations where racks or mangers are available. Conventional sized bales are particularly convenient to handle and enable rationing to be easily undertaken. Large bales are increasing in popularity but need a specialised manger placed in a convenient position for loader filling, or a long feeding passage where bales can be easily unwrapped. Wastage can be a problem with hay feeding, especially when the quality and palatability are low.

3. Straw

Although straw is widely used in cattle enterprises for bedding and for feed, considerable developments are taking place in the methods of chemically treating the material to improve its digestibility. On-farm treatment of chopped straw with caustic soda is a specialised and potentially hazardous operation which is usually undertaken by contractors. Treatment with ammonia can be undertaken in specialised ovens on a batch system or by pumping the aqueous form into a stack of bales completely sealed with plastic sheets. Chopped or milled and treated material, as well as purchased cubed straw, can be readily included in complete diets and bales used as a part or whole replacement for hay. The use of treated straw for beef cows and replacement heifers is expected to increase considerably in the future. It will no doubt also have an increasingly valuable role in dairy cow rations to optimise fibre levels and assist in the control of fermentation.

CONCENTRATE FEEDING

It was explained in the last chapter that the amount of concentrates to be fed to dairy cows depends upon the quality of the background feeds, the level of milk yield and the appetite of the animals, which is affected by the stage of lactation.

A simple guide to rationing is shown in Table 3.1.

Table 3.1 Rate of concentrate feeding

Milk from background feeds (kg per day)	*Milk yield per day (kg)*				
	10	*15*	*20*	*25*	*30*
	Concentrates per day (kg)				
M	4	6	8	10	12
M + 2.5	3	5	7	9	11
M + 5	2	4	6	8	10
M + 7.5	1	3	5	7	9
M + 10	—	2	4	6	8
M + 12.5	—	1	3	5	7

It is usual to feed all or part of the concentrate ration at milking time so that individual cows can be rationed according to such a scheme as in Table 3.1. Alternative systems of allocating concentrates, however, have been developed to avoid the need for frequent changes of the cows' ration and the associated frequent adjustment of feeders at milking time, thereby simplifying the herdsman's job.

Step Feeding

This is a system developed to simplify feeding and also to encourage high peak yields. Fixed daily amounts of concentrates are fed for predefined periods of the lactation, each usually lasting 8–12 weeks. The level adopted is related to the yield in early lactation (days 10–14) which has been shown to be a good guide to the anticipated potential of the cow, so long as she calved normally and has not been ill. If the herd is sufficiently large to justify subdivision into yield groups, allocation of background feeds and concentrates becomes relatively simple. Facilities to enable the feeding of some concentrates out of the parlour will be necessary for high-yielding animals. It is usual to feed about 9 kg per day in the first two weeks of lactation until potential yield is obtained. Table 3.2 gives examples of rations based on forage of varying quality.

Table 3.2 Daily diets

Predicted peak yield			*35 kg/day*			*27.5 kg/day*	
Predicted lactation:			*7,000 kg*			*5,500 kg*	
Forage quality	*Period no.**	2	3	4	2	3	4
Very high	Forage DM (*kg*)	12.5	14.0	15.5	15.5	16.0	15.2
	Concentrate (*kg*)	7.5	4.0	2.0	1.0	0	0
Medium	Forage DM (*kg*)	7.0	7.5	8.0	8.0	8.5	8.5
	Concentrate (*kg*)	13.5	11.5	10.0	9.5	8.5	7.5

* Period 1 = 0–2 weeks; 2 = 3–12 weeks; 3 = 12–22 weeks; 4 = 22 weeks
Source: Adapted from MAFF.

Care should be taken to avoid reducing the plane of nutrition around the time of conception, particularly in a high-yielding animal which has had excessive liveweight loss in early lactation.

Flat-rate Feeding

This is another system which avoids concentrate rationing of individual cows. It is a simple system in which a single level of concentrates is fed to the whole herd or to a group within the herd. High-quality forage (usually silage) is fed ad-lib so that higher-yielding cows due to their larger appetite can eat more silage.

The system is based on the fact that body weight and appetite vary through lactation and that reducing the level of starchy feeds in the diet increases the intake of high-quality forage.

Yield potential is again predicted (10–14 days) to enable an appropriate level of concentrate feed to be selected for the whole of the winter feeding period.

A guide to winter feed requirements for flat-rate feeding is given in Table 3.3.

Table 3.3 Winter feed requirements for flat-rate feeding

Breed	*Calving season*	*Forecast yield (kg)*	*Daily concs. (kg)*	*Likely daily silage intake (kg)*	*Winter feed (tonnes) Conc.*	*Silage*
Channel Islands	Sept/Oct	5,000	7	31	1.37	7.0
		4,000	5	31	0.98	7.0
	Jan/Feb	4,500	7.5	35	0.55	5.5
		3,500	5.5	33	0.40	5.3
British Friesian	Sept/Oct	6,500	8	38	1.56	8.6
		5,500	7	35	1.37	8.0
	Jan/Feb	6,000	8	46	0.58	7.4
		5,000	7	43	0.51	7.1
Holstein	Sept/Oct	7,000	9	39	1.76	8.8
		6,000	7.5	37	1.46	8.4
	Jan/Feb	6,250	9	46	0.66	7.5
		5,250	7.5	43	0.55	7.3

Source: Adapted from MAFF.

High protein concentrates On farms where adequate quantities of high D value silage is available (grass or maize), high feed intakes can be achieved by supplementation. For example, a diet comprising 55 kg silage (11 MJ/kg DM) + 5 kg concentrates (13.0 MJ/kg DM) can produce 30 litres of milk (equivalent to 0.15 kg conc/litre).

Complete-Diet Feeding (CDF)

Now used with some 10% of UK cattle, the CDF system involves supplying the animals on an ad-lib basis with a uniform mix of chopped or coarsely milled forage, concentrates, cereals, minerals and other feeds. The mix is prepared and distributed using a mixer wagon which is normally fitted with load cells so that accurate weighing of the ingredients can take place. The herd, which has to be sufficiently large to justify the capital cost of the equipment, is subdivided into yield groups with diets prepared with differing energy densities according to the requirements of the average animals in each group. A small proportion of the herd may become over-fat but are valuable culls.

Table 3.4 Ration formulation

Stage of lactation	*Approx. DMI (kg/day)*	*Suggested M/D (MJ/kg DM)*	*Crude protein %*	*Forage DM % total DM*
Early	19.0	11.5	16	35–45
Mid	16.5	10.8	14	55–65
Late	11.0	9.8	12	70–80
Dry	11.0	9.0	12	80–90

Many farmers using CDF have discontinued the feeding of concentrates in the parlour although others feed a small quantity in order to encourage reluctant cows to enter for milking. The system is capital intensive, and can be efficient in labour use although the 'feeder' does require a high level of skills in stockmanship, mechanisation and even mathematics! The time taken to load, mix and distribute the diets depends upon such factors as building layout and distance between units, if, for example, rations are also similarly prepared for youngstock and beef cattle. Location and layout of forage silos and other ingredient stores are, however, major factors in efficient labour use. Brewer's grains and sugar beet pulp, as well as chopped straw, are common ingredients of a complete diet or total mixed ration (TMR) as it is sometimes called, so these should ideally be stored near the silage clamps. Economic benefit is frequently achieved by the concentrate, particularly from the protein proportion of the mix provided by purchased 'straights'. These are usually in the form of meals delivered in bags or bulk and are derived from fish, soya, rape-seed or maize gluten. Energy-supplementing ingredients are also available, including such materials as molasses (for which a suitable high-level storage tank is required), cereals and specially formulated oil and fat products.

CDF is a system which offers considerable scope on the one hand to increase milk yield and compositional quality, and on the other to minimise concentrate cost by the purchase and inclusion of good 'buys'. Careful planning of diets is essential, as is a gradual rather than abrupt change of the mix. The secret of success, as in so many areas of farming, is to carefully control the whole procedure, ensuring the viable operation of the mixer wagon.

Out-of-Parlour Feeders

This is another capital-intensive system of feeding concentrates but one which is being increasingly utilised in medium-sized, high-yielding herds where CDF is not feasible.

It is a system which fits well with the self-feeding of silage, especially in building layouts where it is inconvenient to locate mangers for concentrate feeding. The system involves a number of feeding stations (each serving 25–30 cows), which are best sited between the forage feeding area and the cubicles or bedded yards.

Cows are fitted with transponders on a neck collar and after a period of training soon get used to entering the stalls of the feeder and receiving their ration of concentrates. The current type of system being installed rations the cows individually, using a microcomputer which also displays the cow's daily consumption as well as the total daily usage. The system can be of help to herdsmen by displaying the number of an animal which has not eaten her allocation because she may well be 'off-colour' or in oestrus. As with other feed dispensing equipment, calibration needs to be frequently checked.

Manger Feeding

Concentrate feeds over and above that received in the parlour at milking times can be fed to high-yielding cows in an outside manger or at a feed fence. Bagged food is a convenient method of handling and certainly of rationing, although various types of trolley and mechanical dispenser can be used. A facility to be able to yoke the animal is ideal; it is a high capital cost but prevents

An out-of-parlour feeding unit with three stalls and integral bulk-storage hopper.

Cows using the out-of-parlour feeders. The spring-loaded rear gate prevents other cows entering the stall whilst an animal is using the feeder.

bullying and can also be useful to herdsmen while undertaking many of their husbandry tasks, even for AI. When feeding at a manger without yokes, the ideal feeding space for Friesian type cattle is 65–70 cm. This amount of space will require the cattle to pack together so that they are less likely to back-out knowing that they will have difficulty finding an alternative place along the line. The provision of a manger feed late at night is a good method of increasing nutrient intake but disturbance to the herd at this time makes the task of oestrus detection much more difficult.

CHAPTER 4

SUMMER FEEDING

Herdsmen, with few exceptions, look forward to the spring months when the cattle can be turned out to pasture and the work load associated with winter chores is reduced. As grazed grass forms the major part of the diet of beef and dairy cattle during the summer months, it is important for herdsmen to be aware of the factors which affect grass growth and its utilisation by the stock.

Following the relatively precise feeding systems of the winter comes the difficulty of dealing with a material which is highly variable in terms of both quality and availability. The reasons for this variability are dealt with in this chapter, together with the practices which can be undertaken to get the most out of what is the cheapest type of feed. Before and during the season many important decisions have to be taken such as which fields to graze, cut or fertilise. In many of these situations the decision will have to be taken by the herdsman alone, but in others, it will be the management, hopefully in consultation with the herdsman, who decides on a particular course of action.

Factors Determining Grass Growth

Before dealing with the alternative systems of grazing and the factors which influence efficient utilisation, it will be helpful to consider a number of factors which determine grass growth.

Farm Location Although temperature and radiation from the sun influence growth, the most important factor in Britain is the rainfall during the growing season (April to September). Soil type is also a key factor, especially its water-holding capacity. The two factors of rainfall and soil type can be used to predict the grass-growing potential of a particular farm or field. A system of classification has been worked out at the Grassland Research Institute based on site classes from 1 (very good) to 5 (poor). By reference to Table 4.1 the grass growing ability of a particular location can be obtained.

Table 4.1 Grass growing site classes

	Average rainfall (April–Sept)		
Soil texture	*More than 400 mm (16 in)*	*300–400 mm (12–16 in)*	*Less than 300 mm (12 in)*
Clay loams and heavy soils	1	2	3
Loams, medium textured soils; deep soils over chalk	2	3	4
Shallow soils over chalk or rock; gravelly and coarse sandy soils	3	4	5

The total dry matter yields from each site class from three cuts and a subsequent grazing were:

Site class	*DM yield (t/ha)*
1	12.7
2	11.6
3	10.5
4	9.5
5	8.4

It is interesting to note in Figure 5 that most of the difference in yield occurs after June, when poor sites suffer most from mid-summer moisture stress.

Types of Grass The species of grass being grown does not have such an important effect on yield as many farmers and stockmen think. Ryegrass, timothy, meadow fescue and clover are certainly preferred species but good yields of digestible forage can be obtained from permanent swards which have a high proportion of useful indigenous grasses. Over half the lowland grass used by dairy cows in the United Kingdom is unsown or is more than ten years old. Apart from the swards with a high weed population it is usually worthwhile improving and maintaining pasture quality by management rather than incurring the high cost of reseeding.

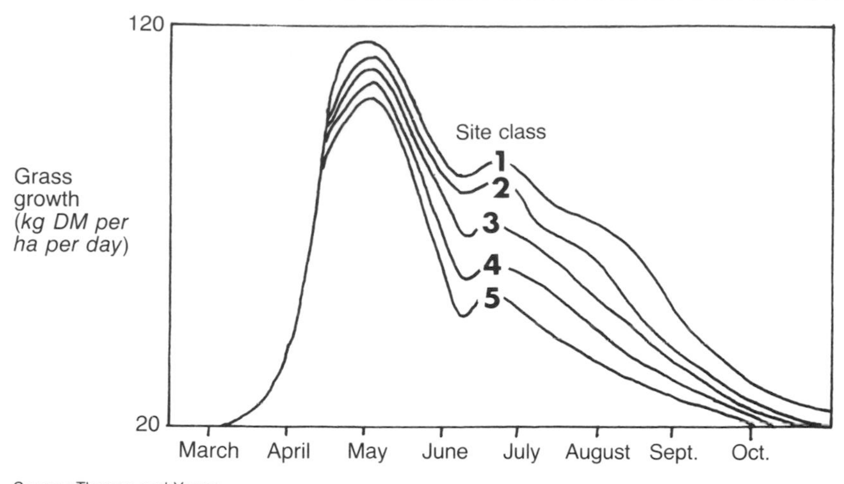

Source: Thomas and Young.

Figure 5. Seasonal pattern of yield of S23 perennial ryegrass.

Sown grasses do provide earlier spring growth (especially Italian ryegrass) and heavier silage cuts, and offer less risk of parasites to grazing stock. Permanent pasture, however, is more drought resistant, withstands treading much better, especially in high rainfall areas, and is well suited to small intensive farms where the risk of a failed ley can be a serious matter.

Clover has higher intake characteristics than grass and saves on nitrogen fertiliser costs. It does, however, contribute little to output in the early part of the season or on farms where more than 200 kg/ha N is used. With the increasing interest in 'green' issues and particularly in organic meat and milk production systems, clover swards become invaluable.

There is an increasing but still relatively small demand for organic milk. Conversion of land to organic standards set by the United Kingdom Register of Food Standards (UKROFS) takes two years, although livestock can be converted in three months. Similar standards for welfare and housing apply as with normal production but although drugs can be used to prevent animal suffering, long milk withdrawal periods operate. A 20% premium on milk price is required to compensate for reduced output, primarily due to the lower stocking rate.

Lime status The pH of the soil should be monitored by regular tests and lime applied to maintain a level of 6.0. The better species tend to die out if the soil is allowed to become too acidic and the response to fertiliser is reduced. The more nitrogen is used, the more frequently lime will be required.

Weed control Docks, thistles and buttercups are common problems in permanent grass as are arable weeds in many newly established leys. MCPA and 2,4-D are suitable chemicals to use for control where no clover is present; otherwise MCPB and 2,4-DB should be used. (Docks may need several treatments.)

Phosphate and potash Grazing allows much of the P and K to be returned to the soil so that on fields which have been intensively stocked, annual dressings of 30 kg/ha P_2O_5 and 40 kg/ha K_2O are usually adequate. When fields are frequently cut for conservation, increased K is needed at a level of 0.5 kg K_2O for each kilogram of N applied. Soil analysis should be undertaken to monitor P and K levels. If slurry is applied the above levels can be reduced or even eliminated. Large single dressings of K should be avoided; otherwise 'luxury' uptake occurs which is not only wasteful but also increases the risk of hypomagnesaemia.

Nitrogen Because this is the major nutrient affecting grass growth, its usage has a marked effect upon the productivity of a cattle enterprise. Although the average application on UK dairy farms is little over 150 kg/ha, the economic optimum or target levels vary between 300 and 450 kg/ha depending upon site class. See Table 4.2.

Table 4.2 Target levels of nitrogen use

Site class	*kg/ha*
1	450
2	400
3	350
4	325
5	300

The distribution of yield through the season can be influenced by the timing of applications. Heavier mid-season applications will improve production at this time so long as adequate soil moisture is available. The first application of the season should ideally take place before growth commences, with subsequent dressings to grazed areas every 3–4 weeks. The level applied should be 1.7 to 2.5 kg/ha per day depending upon site class. For silage cuts, 120–150 kg/ha is recommended for the first one and 100 kg/ha for subsequent cuts. Special formulations of nitrogen and phosphorus encourage early growth, produce higher yields of first-cut silage and minimise any risk of leaching through the soil.

End of season management There is a tendency for grass, especially in sown leys, to continue growing well into the autumn when the economic utilisation by dairy cows becomes infeasible. Leaving this uneaten material in situ can cause problems, especially in a hard winter when a high proportion of the plants can be killed out. The ideal management tool is a flock of grazing sheep—although not all dairy farms have access to such a facility—or appropriate fencing. Health risks from such 'imported' sheep need to be watched, particularly in respect of leptospirosis. Grazing with young stock may be a possible solution but to maintain adequate growth rates at this time, supplementary feeding will almost certainly be required. Any animal used to remove surplus grass should be taken off the area before Christmas ideally, sooner if very wet conditions occur.

Forage Quality

Having grown the grass, then comes the challenging task of utilising it efficiently for grazing. The first factor is that quality varies markedly throughout the season. This is measured in terms of digestibility of the material (D) which is the percentage digested by the animal. Young, very leafy material in early spring can be as high as 80 D so that supplementation with a fibrous feed such as hay or straw is essential; otherwise it moves through the gut too quickly and inefficient digestion takes place. The aim should be to present the animal with material that has a good proportion of leaf to stem and rarely falls below 70 D. The D value as well as level of crude protein falls as the grass begins to flower, with a mature crop having a digestibility of below 60 D. Each variety has a characteristic relationship between digestibility and stage of growth, e.g. Timothy declines well before ear emergence. Autumn grass is not digested as efficiently as spring grass due to the lower sugar levels (young spring stems have a higher sugar content than leaf).

Regular topping of pastures removes any seed heads and

ensures that higher quality material is available at the next grazing. This is a suitable job for herdsmen to undertake; it not only gets them out of the buildings into the 'fresh air' but also allows them to tidy up the grazing, which can be good for morale. If the cows are in the next field or paddock it is often possible, whilst driving up and down, to observe their behaviour over an extended period.

Grass which becomes soiled by urine or dung is unpalatable for several weeks and tends not to be eaten. This problem appears to be more serious when stocking rates are at an average level rather than when low or very high (presumably the cows get used to soiled grass). Unless heavy rain soon follows, harrowing of dung pats can make the situation worse.

Amount of Grass on Offer

Thick, dense swards of reasonable length allow the grazing animals to consume a full diet with minimum movement and exertion. Sparse, over-grazed crops, however, require the cattle to work harder and after several hours they become fatigued and discontinue grazing, so that appetite is not met. Production is then seriously affected as maintenance requirement is increased at a time when nutritional intake is low. Grass consumption on a good sward can be as high as 70 kg per day but is only half that amount on a thin poor one. Growth of grass can be erratic due to rapid weather changes, so it is an important job for the herdsman to check on the amount of grass available. On wet and windy days the animals will spend time sheltering to the detriment of grazing, and when they do graze, dry matter levels will be low. A useful task is to measure the length of grass after grazing with a ruler or perhaps using marks on the wellington boot! Ideally, when the cattle leave a rotationally grazed area, the average height of remaining grass should be between 8 and 10 cm (about 4 in). In set-stocked areas, production will be reduced if the average height of grass falls below 6–8 cm (3 in).

Stocking Rate

This is another key factor affecting the profitability of grassland but one which can be agreed and set before the grazing season begins. Low stocking rates lead to uneaten grass which dies and is mainly lost. Remaining material becomes stemmy, digestibility falls so that intake and milk yields also fall. Too high a stocking rate leads to overgrazing and also to the lowering of intake and performance.

Grazing stocking rates need to fall during the season to match the natural decline in growth and quality of grass. On a class 1 site, 6 cows per hectare can be carried up to mid-May but only 2.5 in the late season. Areas which have been earlier cut can be added to the grazing area as the season progresses. Conservation is not only a means of producing stocks of winter feed but is a vital tool in the management of summer grazing.

Irrigation is a technique which a few farmers are able to use to improve stocking rates or, in other words, reduce the risk of a high stocking rate in dry years. It is in many ways a method of improving the site classification, but it is usually obtained at high capital cost and on some farms, which still use hand-moved pipes, has a high labour requirement. Even on farms equipped with automated, self-wrapping hose reels, it is often the task of the herdsman to move the equipment once or twice a day to its next 'run'. This is yet another job which needs to be planned with care; otherwise it will most probably clash with milking, or even supper time!

Shortage of fodder is a frequent problem for herdsmen aiming to obtain respectable stocking rates on poor grass growing farms, but one facing most at some time. It can be dealt with by the purchase of feeds such as brewer's grains or arable by-products. Other farms buffer the grass with silage, hay or straw, and in fact this system of supplementing grazing is becoming more common as a routine procedure to maintain or even boost milk yields and compositional quality. The availability of silage in big bales, together with the wrapping technique and ring feeders, has produced an efficient and easy system of buffer feeding. Maize silage is a good 'balancer' to grass, but care in managing the clamp face in the summer to avoid secondary fermentation is essential.

Increased concentrate feeding is frequently involved but this subject will be discussed below. Lucerne with its deep rooting system can be useful when there is a shortage of summer fodder, as it continues to grow during dry periods but it can only be reliably grown on free draining soils of high pH in the southern and eastern counties. Bloat is a potential problem if lucerne is grazed; this can be minimised by cutting, wilting and then grazing with control from an electric fence. Rape and early planted kale are other alternative feeds which can supplement grass in the summer and early autumn months. Germination and growth of these crops is also highly weather sensitive, so that failures are not unknown.

The Milk Potential from Grass

It has been explained that when grass of the appropriate quality and quantity is available, the cow will satisfy her appetite and produce to her potential. The situation can arise in which the cows are yielding less than the amount that the grass is capable of supporting. Unless stocking rates are increased accordingly, grass in these situations is wasted. In herds which have a wide spread of calving dates, the difficulty arises in balancing the needs of individual animals with stocking rate and with fertiliser use. Subdivision of a herd into yield groups does help this situation but at the same time it increases the time spent by stockmen in bringing cows to and from pasture. The potential milk production from high-quality grazing is shown in Figure 6.

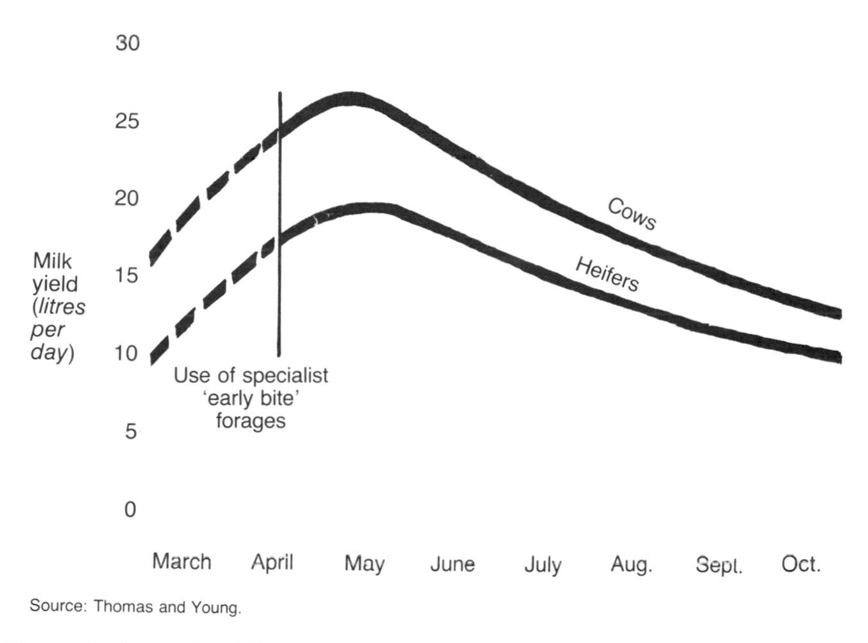

Figure 6. Potential daily milk production for ad-libitum high-quality grazing.

Supplementary Concentrate Feeding

The feeding of concentrates to dairy cattle during the summer months, although practised by many producers, is seldom a profitable activity. It may be justified for short periods when grass growth is slow, but when adequate forage is available the animal substitutes concentrates for grass and produces very little extra milk. When grass is in short supply, body condition can soon be lost, resulting in a long-term effect on cow performance, so that this is another situation in which concentrate feeding is justified.

High-yielding spring calvers will also justify concentrate feeding if their yield level is above the potential of the grass. This will be more so with fresh calvers whose appetite is still at a low level. There is no fixed rule for the rate of feeding concentrates in the summer but the type of concentrate to be fed is important. The protein in grass is highly degradable, particularly in mid and late season, so that high-yielders require undegradable protein supplementation. Energy is usually limiting for high-yielders and also in order to maintain adequate butter fat levels, the type and content of the oil in the diet have to be taken into consideration. In terms of minerals, grass is usually lacking in phosphorus, and in some cases in trace elements, and, of course, magnesium.

The general conclusion about the feeding of concentrates at grass is that the herdsman should have confidence in the grass, give it a chance to produce the milk and supplement with concentrates only at critical times. The rate of decline of milk yield, as well as the body condition of the animal, should be regularly monitored and buffer feeding introduced, ideally from forage type feeds, if there is any sign that grass is not fully meeting requirements. It may even be a sensible precaution to have some buffer feed on offer, i.e. a rack of hay or straw, throughout the season at some convenient place in the dispersal area.

GRAZING SYSTEMS

Several methods of controlled grazing are in use for dairy cattle, all of which aim to provide the animals with adequate forage but at the same time minimise wastage.

1. Strip Grazing

This system involves the use of an electrified fence which is moved once or twice daily to provide the herd with a clean strip of forage. It is an efficient system of rationing and generally avoids wastage, although a back fence is necessary in a large field to prevent eating of the re-growth. Although the system is rather labour consuming for herdsmen, the biggest problem is usually the one of deciding how far forward to move the fence. Providing too much grass causes selective grazing and wastage but too little fails to meet appetite and the cows are also more likely to break through the fence! In wet conditions on heavy soils, poaching can be a problem, especially when the system is used to ration kale or roots. Surplus grass can be taken more easily for conservation than with other systems and the provision of water troughs seldom causes a problem.

There is some skill involved in putting up an electric fence efficiently. Ensure that the end posts are firmly in place and then have the wire as straight and tight as possible. The height of the wire, which is usually set by the use of 'standard' pig-tail posts, is normally 80–90 cm from the ground. Mains supply fences should be installed and maintained according to the manufacturer's instructions and batteries replaced in portable units, as required.

2. Paddock Grazing

With this system, permanent and semi-permanent fences are used to subdivide the grazing area into a number of paddocks, so reducing the labour requirement for moving fences during the summer. The system is based upon graze, fertilise and rest for a period until grazed again. The interval between grazings may be as short as 20 days in the early part of the season, extended to over 40 days in the autumn. At the peak of growth, surplus paddocks can be cut but problems do arise with large machinery operating in small and awkward shaped paddocks. When laying out the fences it is worthwhile arranging for the paddocks to have parallel sides which make fertiliser application, harrowing, topping and irrigation much more efficient. Ideally, the width of each paddock should be multiples of twice the working width of the fertiliser applicator (to avoid empty runs back to the gate).

Some farms operate with many paddocks, whereas others have a few larger ones and some even strip-graze within the paddocks. Daily milk yield can vary considerably when using a few large paddocks, this being an example where access to a buffer feed minimises such problems. Adequate water should always be available, perhaps one trough in the fence line shared between two paddocks. Good roadways are advantageous, not only for vehicular traffic, but also for the efficient movement of the herd. This system has lost popularity due primarily to the high cost of fencing and water supplies.

3. Set-stocking

This is the traditional system of grazing which allows the cattle free access over a set area of pasture. It now carries stocking rates almost as high as in rotational grazing. Nitrogen is applied in blocks or over all the area, usually on a 3–4-week cycle. The size of the area can be adjusted after each silage cut and another form of stock, e.g. sheep and lambs, drafted in for a short period if undergrazing is a problem. To ensure that the grass does not initially 'get away' from the cattle, turn-out has to be early, with buffering by winter rations until grass growth is balanced with herd need.

One of the major limitations to this system for herdsmen is the time involved collecting the cows, especially with the larger herds in big fields. The problem can be eased by operating with two fields, one for day use and the second more convenient one for night grazing.

Milk production on each of the three systems is very similar except that rotational systems do have a slight edge over set-stocking for high-yielding spring calvers. The system selected for a particular herd will thus depend upon farm layout or, more usually, upon the preference of the people involved.

Water Supplies at Pasture

Ideally every field or paddock should have a water trough and it should be sufficiently large so that at least 10% of the group can drink at any one time. The average consumption varies according to the weather conditions and milk yield, but on a hot summer's day high-yielders can drink as much as 180 litres each. Cows are reluctant to walk more than 300 metres to drink. On farms where herd size has increased dramatically it will usually be necessary to increase the diameter of the supply pipe from the customary 12.7 mm (½ in.) to 19 mm (¾ in.) or even 25.4 mm (1 in.). The provision of a concrete apron may well be justified where heavy soils and frequent use cause a seriously poached area. Cow comfort is improved and a reduction in foot problems would be expected from such an investment.

Collecting cows from paddock grazing. This is an easier task for the herdsman than collecting from a set-stocked area.

A large-capacity water trough, allowing at least 10 per cent of the herd to drink at any one time.

Zero-grazing

A small number of milk producers—where there is an unhandy farm layout or perhaps where a motorway has subdivided the farm—cut and cart fresh grass to the herd throughout the summer months. Improved levels of utilisation occur due to the elimination of selection and fouling, so that higher stocking rates are achieved. There is no need for fencing or water supply but the available mechanical skills do need to be considerable. This system can create considerable stress on cows and machines, as well as on herdsmen and management, particularly if the slurry system has difficulty in coping.

Storage Feeding

Using conserved feeds as the basis of feeding throughout the year allows for high stocking rates and avoids the daily chore of cutting and carting grass. It is a system widely practised in such countries as the USA and Israel, where regular supplies of high-quality feeds (e.g. lucerne hay) can be purchased economically. It is now being used in the UK by only a few producers, but some are showing interest because of the increased control which is possible over feeding and milk production. Cow health is an obvious potential problem as is machinery reliability, but it will no doubt be a system more widely used in the future.

A mixed system could have more attraction to herdsmen, i.e. storage feeding by night and grazing by day. The cows are then handy for morning milking, which is especially valuable when they are milked three times a day. Temporary grass shortages are easily compensated by the animals increasing their silage (or hay) intake at night. Increased silage quantities are obviously involved but it is a system offering considerable flexibility in that it can be readily introduced or discontinued according to silage/grass availability. This is really an extension of buffer feeding as described above.

Folding Green Crops

The strip grazing of kale, roots or even sugar beet tops is a method of extending the grazing season. With kale it is preferable to provide only two or three rows rather than a deeper strip, so as to minimise wastage. An alternative system is to cut and cart the kale, ideally using an offset, side-mounted forager to minimise soil contamination.

When grazing green crops it is also preferable to provide a grass or stubble lie-back area and to avoid long walks to and from the field. There is some risk with digestive upsets in frosty weather, so that a hay feed should be given before cattle are allowed to graze kale.

Forage Utilisation

It has been shown that the main aim in summer feeding is to make the full use of pasture in conjunction with other supplementary feeds. A simple estimate of the energy contribution of grass and forage on a farm is contained in the concept of utilised metabolisable energy (UME). This figure is obtained by subtracting the energy supplied by the concentrates and other purchased feeds from the total energy required by the animal. The UME of a farm with well-managed grassland is 75–90 GJ per hectare (1 GJ is equal to 1,000 MJ). Some dairy recording and costings schemes calculate the output of milk per hectare obtained from forage (a good target being 9,000 litres per hectare).

CHAPTER 5

COW MANAGEMENT

Having dealt with the important topic of feeding, we now move to an equally important one for herdsmen—that of managing the cows through the production cycle. Management is a word used in several contexts but what is meant in this situation is the provision of the best possible working conditions for the cattle. This is the day-to-day tactical or operational management rather than the business or financial management in which the herd owner or manager is more likely to be concerned.

Herd management, as with any management task, involves planning, making decisions and control, but it also involves herdsmen taking action at the right time, without unnecessary cost, to ensure that the cows are healthy, well fed and producing at the optimum level. It will usually involve using the resources that are available such as the buildings, feed supplies, other staff and machinery, but it may from time to time require bringing in extra resources such as the vet or a freeze-branding contractor. Many tasks will ideally have been planned well ahead and agreed with the owner; others, however, will be unexpected (a cow calving three weeks early) or a real emergency (several bloated cows). Herdsmen need to know how to cope in all such situations, and this will come with experience helped (hopefully) by reading these pages!

The varying needs of the dairy cow through the production cycle will be considered in this chapter, but first we will cover several techniques involved in efficient herd management, i.e. identification, herd grouping and body condition scoring.

Identification

All cattle, but especially dairy cows and youngstock, need to be clearly marked for easy and accurate identification. For health control reasons there is a statutory requirement that all cattle are permanently identified with an approved type of metal tag or ear

A metal ear-tag being placed in the right ear of a calf, as required before 14 days of age.

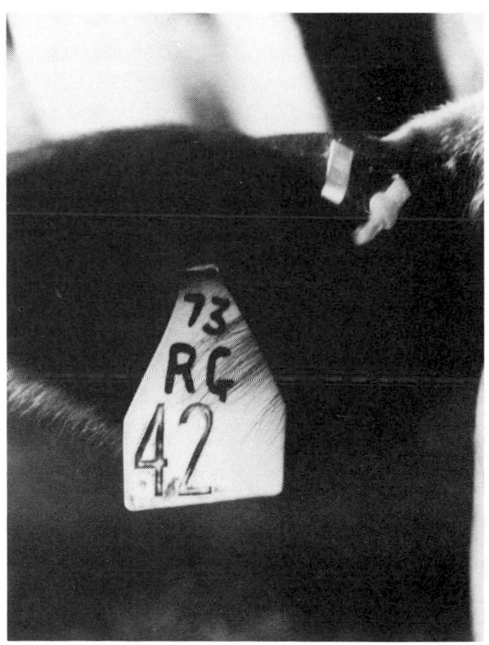

A plastic ear-tag carrying the number of the animal (42), the initials of the sire (R.G.) and the year of birth (1973).

tattoo. Each herd has a mark allocated by the Divisional Veterinary Officer in a letter–number combination on a county basis. The ear-tag has to be placed in the right ear of the calf before it is 14 days of age.

Large plastic ear-tags can be a most satisfactory method of identifying calves and youngstock until they are of a suitable size to be permanently marked, for example by freeze branding. Even with adult cattle, a large ear-tag can be helpful when observing the animals from in front, as at a feed fence. Interest in the subject of herd breeding is often aided by adding to the tag the initials of the animal's sire as well as the year of birth. Different coloured tags for each breeding season or progeny group can be used to identify 'similar' animals.

Freeze branding is now the most common method of identifying cattle, especially for black and white cattle, this being preferred to plastic collars, neck chains or ankle straps. The system of milking determines the site of the marks, but for herringbone milking the rear of the flank is usual (but on a black patch). Freeze branding is a patented technique so it should only be carried out by the licensees of the patent holder. The pigment-forming cells of the hair follicles are destroyed by the low temperature of the branding iron, and the subsequent growth of hair is white—hopefully for the lifetime of the animal. Increasingly, branding is being carried out at the yearling stage so that the numbers are ready for use during the first service period.

The trend towards electronics in dairying involves the use of automatic identification with transponders attached to a neck collar. The energising and interrogation equipment is located at the stall in the milking parlour, at the out-of-parlour feeding stations or at the walk-through weighbridge. One can imagine that before too many years it will be commonplace for animals to have a permanent lifetime transponder 'implanted' into the rumen. Such a device should be a deterrent to cattle rustling and be of value to a national data collection scheme if and when one is established. There is every likelihood that such a unique coding system will be introduced in the near future to aid management as above but also for MAFF to monitor animal movement and identify, for example, offspring of BSE cases.

Temporary marking to identify the cows which are undergoing antibiotic treatment or require drying-off can be achieved by using coloured grease crayons or aerosol sprays. Adhesive tape can also be used on the tail for differential feeding or other management purposes, for instance a cow to be culled, so as not to be

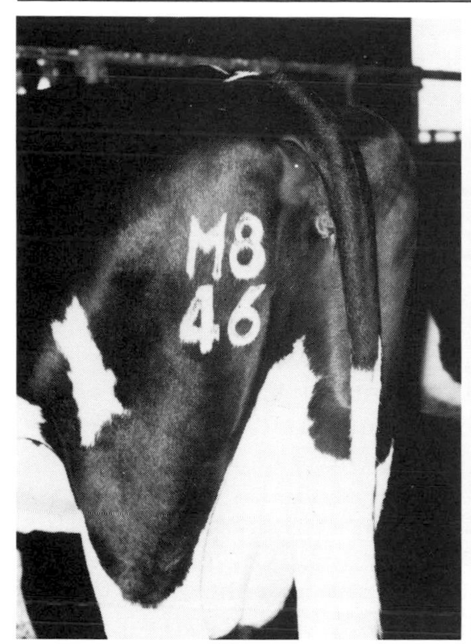

A freeze-brand mark, ideal for general identification but especially at milking time.

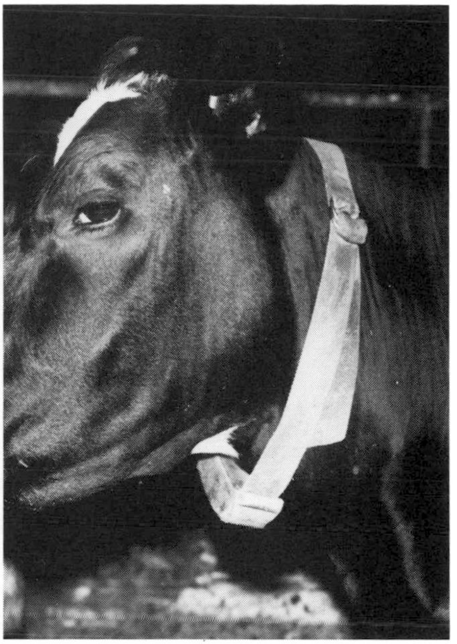

A transponder fitted to a neck collar for automatic identification.

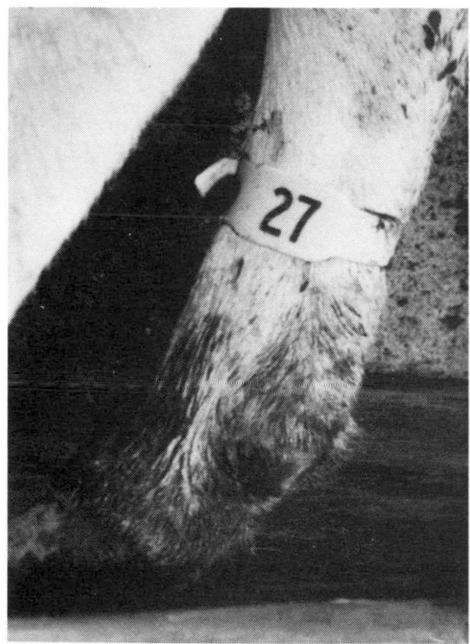

Ankle strap.

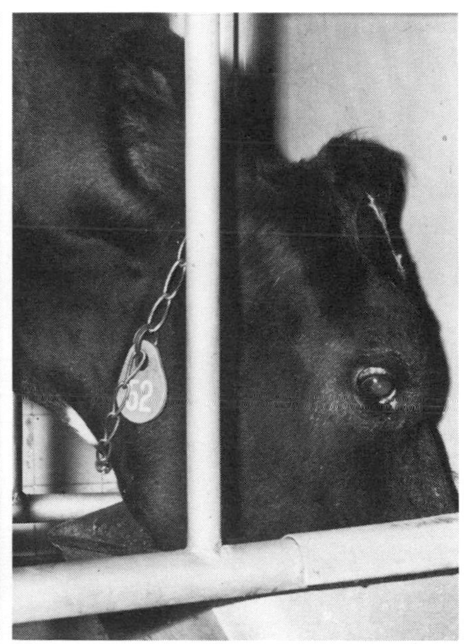

Neck chain.

re-bred. The major problem with such tail tapes is that if they are fitted too tightly, the blood supply to the switch is affected with serious results.

Grouping of Cows

Reference has already been made to the benefit of subdividing a herd into groups for differential feeding and management purposes. With cowshed housing the ultimate in individual feeding is possible, as it is with yokes or individually programmed out-of-parlour feeders. In loose-housing situations grouping has much to offer, especially in herds of over 80–90 cows and when rationing of concentrates in the parlour has been replaced by a block-feeding scheme such as complete diet feeding or step feeding.

Grouping allows one or more dry groups to be formed, perhaps one of the first lactation heifers as well as subdivisions of the main milking herd. The number of groups formed in a specific situation will depend upon such factors as herd size, calving spread, building layout and, especially, the feeding arrangements. A 150 cow herd with a fairly tight calving pattern could have two groups of 75; a herd of 300 could have five groups of 60. As explained in Chapter 1, less bullying appears to occur if group size is kept below 80 but the optimum number should be in multiples of

Subdivision of a herd into groups for feeding and management purposes.

parlour capacity (or of the sides of a herringbone), e.g. 64 or 72 for an eight-stall-sided parlour.

The most debatable point is whether to group by milk yield or date of calving. In the latter situation cows stay in their group, whereas in the former it is necessary to move animals between groups from time to time. Opinion is equally divided as to whether cows lose milk when moved, but in most situations this is to be expected as the level of feeding is lower for the animals moved 'down'. The author has found few problems in moving so long as several cows are moved together and if moving is arranged at a time of low herd activity (late afternoon) rather than at times of high competitive activity (morning feed time). In cubicle houses it is a considerable advantage if there is a passageway from feed to rest areas at each end of the buildings. This allows newly introduced cows to keep on the move and not be trapped in a 'dead' corner, as is frequently the case if only one passageway is available.

Some herdsmen (and owners) consider the extra time involved in moving groups to and from the parlour, especially in the summer months, to be unjustified. Before readily dismissing the technique they are strongly recommended to give it a try, particularly in winter and with first lactation animals, and evaluate the benefits, especially those of reduced stress on the cow. It may well take those few extra minutes to change groups, but this time out of the pit is well spent, having a short break from the milking routine. Morale can be lifted considerably by, for example, washing down the standings after one group and starting the next group—which has a different level of production. Milking need not be a job just to be rushed, and if better oestrus detection is achieved with smaller groups, the extra time involved in changing groups may be very well spent.

Body Condition Scoring

Reference has already been made to the need for herdsmen to monitor the body condition of their cows during the year. Body condition scoring has been developed as a standardised method of assessing the body reserves of the live animal. It is a very useful technique for herdsmen in controlling feeding and in management of the herd, so that it is one with which they should become familiar. This section is written to help the understanding of what is involved; some training and regular supervised practice will also be required before a herdsman can become proficient.

Body condition scoring. The level of fat cover around the tailhead is being assessed.

The level of fat cover over the muscles is what is being assessed, so it is necessary to be able to distinguish between the firm feel of muscles and the softer feel of fatty tissue. The ideal sites for assessment are the loin area between the last rib and the hip bone, as well as on each side of the tailhead. With animals in poor condition the loin should be used but with fatter animals a more accurate assessment can be obtained from the tailhead.

It is necessary to have access to the side and rear of the relaxed

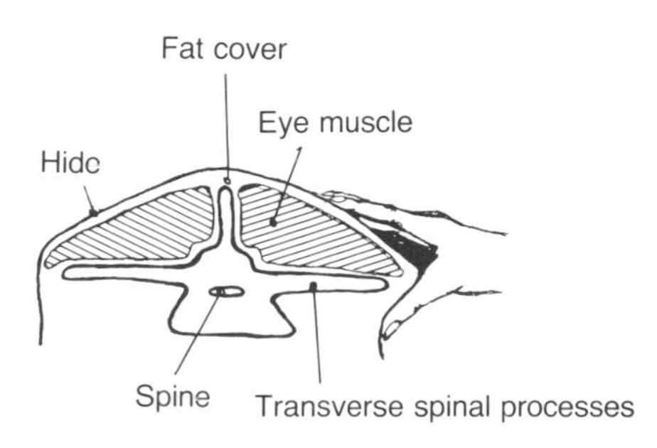

Figure 7. Assessing condition using the loin region.

animal so that the ideal location for the job is in the cowshed, crush or AI stall. On the loin the degree of prominence of the spinal processes is obtained by feeling along the backbone and also by placing the thumb on the transverse process as shown in Figure 7.

Around the tailhead, the level of fat cover is assessed with the fingers and a score awarded from 1–5, as indicated below, to the nearest half point.

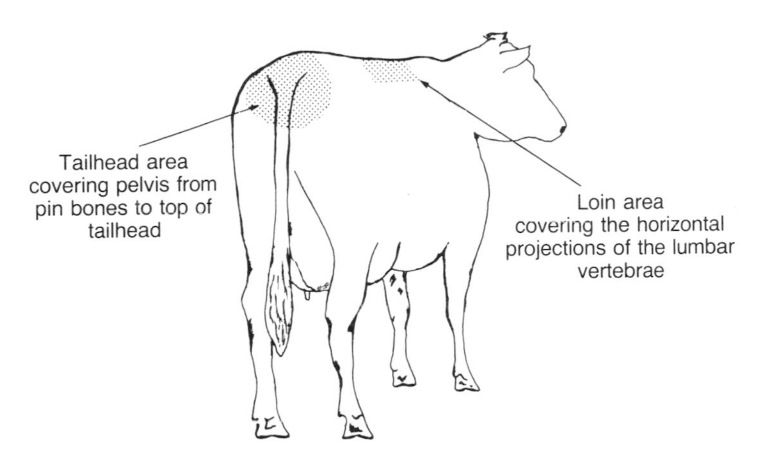

Figure 8. Scoring areas.

If there is a difference in scores between the tailhead and loin areas, it is necessary to adjust the tailhead score accordingly, but by no more than half a point. For example:

Tailhead score	*Loin score*	*Difference*	*Adjustment*	*Adjusted tailhead score*
4	2½	1½	– ½	3½
1½	2½	1	+ ½	2
3	2½	½	0	3

Condition score

1. The individual transverse processes are sharp to the touch and easily distinguished. There is no fat around the tailhead.
2. The transverse processes can be identified individually but feel rounded rather than sharp. There is some cover around the tailhead.
3. The transverse processes can only be felt with firm pressure and the areas on each side of the tailhead have some fat cover.

4. Fat cover around the tailhead is easily seen as slight mounds and is soft to the touch. The transverse processes cannot be felt, even with firm pressure.
5. Very rare in dairy cattle. The tailhead and hip bone are almost covered by fat.

Minimum acceptable condition scores for each month of the productive year of the dairy cow are indicated in Figure 9.

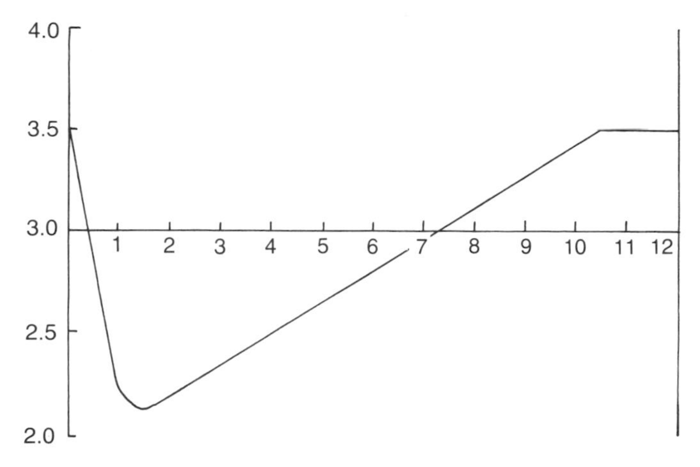

Figure 9. Minimum acceptable condition scores.

COW MANAGEMENT THROUGH THE PRODUCTION CYCLE

Late Lactation

At first thought, it would seem more appropriate to begin a description of the management of a cow through the production cycle by dealing with calving, or perhaps the dry period. However, studies of the results of nutritional research and talks with the owners and stockmen of many 'top' herds, have made it clear that it is in late lactation when the most effective influence on the production cycle can be undertaken. It is for this reason that management in mid lactation will be considered first.

During the second half of lactation, cows should be restored to their ideal body condition (3½ by drying off). For a mature Friesian type cow, this will require a gain of 100 kg comprising: 50 kg for the new calf and uterine fluids and 50 kg to restore the weight loss of early lactation, which in 150 days is 0.66 kg/day. For a young cow which is still growing, a further 50 kg will need to be put on in the last 150 days of lactation, increasing the gain to 1 kg/day. Research work has indicated that it is much more efficient in the use of nutrients to obtain this liveweight gain whilst the cow is milking, rather than when she is dried off. Once the ideal condition is reached it should be maintained by adjusting the feeding rate. Excess body condition is associated with fatty liver problems and is detrimental to milk production, so that it is important not to overfeed even in late lactation. Hence the importance of effective body condition scoring.

Dry Cow Management

The dry period should extend for 40–70 days to allow the udder tissue time to regenerate before the demands of the next lactation. It is preferable to dry off cows abruptly at the end of lactation, treating all quarters with long-acting antibiotics. This treatment controls persisting mastitis infections and prevents most of the new infections that would otherwise occur in the dry period. At the last milking the udder should be milked out fully, the teats thoroughly cleaned with white spirit and the antibiotic infused. The teats should then be dipped in iodophor solution and the udder left to stock up. Feeding for the next few days should aim at minimising milk production by arranging for grazing a bare field or, if yarded, providing straw or poor-quality hay. Cows should be watched closely until the udders contain no milk. Any quarter looking swollen should be checked for hardness and, if there is any, it should be milked out and reinfused with antibiotics. Fly control is important during the summer months so that bi-weekly spraying, or the fitting of repellent ear-tags, should take place. In herds where mastitis is a serious problem or where several freshly calved animals develop infection, it is worth discussing the problem with the vet and perhaps changing the type of dry cow antibiotic. Dry cow therapy is particularly important during late summer, when there is a greater risk of summer mastitis.

Managing dry cows which are in good condition after milk secretion ceases is not a difficult task. The energy concentration of the diet should be kept low—i.e. there should be a high forage intake—in order to keep the rumen extended. It is considered that this is the ideal preparation for a good appetite after calving when an energy-rich diet can then provide maximum nutrients.

Exercise is good for heavily pregnant cows in, for instance, reducing the incidence of swollen and inflamed udders. Grazing is ideal, or if the cows are housed, it is preferable to use a reasonable-sized yard where some regular movement can take place. If individual calving boxes are in use, it is preferable not to move the cow into this facility until very near to calving.

A few cows and heifers may develop swollen and inflamed udders well in advance of calving if 'steaming up' with concentrates takes place at too high a level. This congestion can be relieved by reducing the energy density of the diet, but it may also be necessary to ease some milk from the hard teats twice a day until 12 hours before calving. This milk, which is true colostrum, will be lost as a vital feed for the calf unless it is stored, preferably in a refrigerator. 'Steaming up' with concentrates is a practice which has now been largely dropped, with body condition being maintained by forage feeding. Some dairymen like to introduce a little concentrate (up to 2 kg per day) as calving approaches, but this should have a 'reverse ratio' mineral level, i.e. be low in calcium, to minimise milk fever problems. Magnesium supplementation is also beneficial to the dry cow diet.

Calving

The birth of the calf is a critical time for a cow, and herdsmen need to understand what is involved in the process if they are to provide appropriate care and attention. Although calving is a natural process which ideally takes place unassisted, close observation is required in case difficulties arise. Heifers calving for the first time tend to have more problems than older cows or suckler beef cows, which, unless in-calf to a large breed, usually calve easily.

Animals which calve naturally and expel the after-birth satisfactorily settle better into lactation and re-breed with fewer problems. There is a big temptation for keen but inexperienced herdsmen to interfere too much with calvings, and in most cases to start pulling the calf too soon. This premature interference can easily cause tissue to be torn or bruised, which leads to genital infection and subsequent difficulty in re-breeding. Patience is therefore the number one factor in calving, especially with heifers calving for the first time. There are, however, numerous other factors involved in efficient management of calving which will now be discussed.

It is quite satisfactory, in reasonable weather conditions and especially in daylight hours, to allow the calf to be born at pasture. It can, however, be a time-consuming and even frustrating task for a single-handed herdsman to move cows and newly born calves from field to buildings. Housing in a calving yard or box is more convenient, especially if a complication does arise and the cow has to be restrained for examination and further assistance. It is unwise to calve a cow tied by the neck with a chain. If the cow has to be tied, a halter is preferable, tied with a knot which can easily be released.

The dairy unit should ideally have sufficient loose boxes to allow each cow to calve on its own (see also Chapter 7). The location of the boxes is important to enable single-handed movement from the dry cow yard. Boxes should be sufficiently large to allow several 'helpers' to enter if assistance is needed. Before use, they should be clean, disinfected and liberally bedded with straw. Hygiene is greatly assisted if the walls and floors are rendered, so that effective power washing can take place. An internal hanging gate within the box can be of invaluable aid to restrain and examine a cow about to calve.

The signs of calving approaching are:

- increasing distension of the udder
- stiffening of the teats
- in some cows the 'running' of the milk
- enlargement of the lips of the vulva
- increasing signs of discomfort

The animal tends to detach herself from the remainder of the group and when in the field may select a far from ideal site against a fence or ditch. The best guide to time of calving is to feel for a loosening of the muscles around the tail and for sunken areas between the tail root and each pin bone, which normally appear 12–24 hours before the birth.

Normal Calving

When calving is more imminent the cow is uneasy and often in some pain. She tends to stay on her feet but continues to eat and drink. At this time the uterus is contracting, the water bag pushes through the cervix and the strains get more pronounced. This stage can take 2 to 3 hours with a cow and as long as 6 hours with a heifer.

More vigorous straining indicates the start of the next stage of the process. The calf twists a quarter turn, the head comes through the cervix, the water bag may burst and the feet can usually then be seen. This stage can take up to 4 hours, even

Checking for the imminence of calving as shown by a loosening of the muscles around the tail and the appearance of sunken areas between the tail root and each pin bone. This is usually 12–24 hours before the birth.

longer with a heifer, which tends to tax the patience of an inexperienced herdsman. As the chest comes through the cervix, mucus is squeezed from the mouth and nostrils and this clears the respiratory passages ready for the calf to breathe.

Once the calf is born the cow normally gets to her feet and begins to lick the calf vigorously. The calf staggers to its feet and soon begins to suckle—which helps the afterbirth to be released.

Close observation is required during the birth but so long as some progress (however slow) is being made, there is no need to interfere. The foetus continues to receive nutrients and oxygen via the navel cord until this is ruptured in the final stages of calving. Pulling the calf too soon is the main cause of torn tissue, as the cervix may not then be fully dilated. It is the intermittent pressure of the calf's head which stimulates dilation of the cervix.

If, following several hours' straining, no feet are in evidence, then an examination is justified. This involves scrubbing the hands and arms thoroughly with soap and antiseptic. Lubricating jelly, which many herdsmen find preferable to soap flakes, can be obtained from the vet. It is also necessary to clean the area of the cow's vulva so that dirt is not taken in with the arm. The hand should be pushed in steadily without bursting the water bag. If the head and feet are there, all should be well and more time needs to be allowed before giving further assistance.

It goes without saying that the complete calving kit should be ready and to hand at all times during the calving season. This necessitates washing, disinfection and return to a 'usual' location immediately after use. Stand-by help should always be on call, if not possible from other members of staff, then from friends or neighbours (a reciprocal arrangement having of course been agreed). Readily available telephone numbers (and a phone!) are vital, especially to relief stockmen.

Complications and Malpresentations

It is possible for cows (especially older ones) to suffer from milk fever during calving. One should therefore watch for an unsteady gait, padding of the hind feet and, if in doubt, inject calcium borogluconate. Labour will usually then proceed within half an hour.

If assistance has to be given, e.g. the tongue is beginning to swell, it is best to attach calving ropes to the fetlocks of the calf and pull gently downward towards the cow's hocks, but only when she is straining. If the head has to be roped, it should be placed behind the calf's ears. Sometimes the calf gets stuck at the hips; it is possible to help the situation by turning the cow on to the other side.

When the head is passing through the vulva, the calf's ribs will be passing the cow's pelvis and the umbilical cord is most probably constricted. No time should now be wasted in pulling as the cow strains.

In situations where assistance cannot be arranged, such aids as a calving jack may have to be contemplated but should only be used by experienced stockmen. Before use, it is essential to ensure that the calf is in the correct delivery position.

A common malpresentation is for the head to be tucked back. This problem can usually be corrected by gently pushing the brisket back and, by putting the hand under the jaw, carefully levering the head forward.

Posterior presentation is quite common, i.e. back legs first. This should be recognised by the feet of the calf being upside

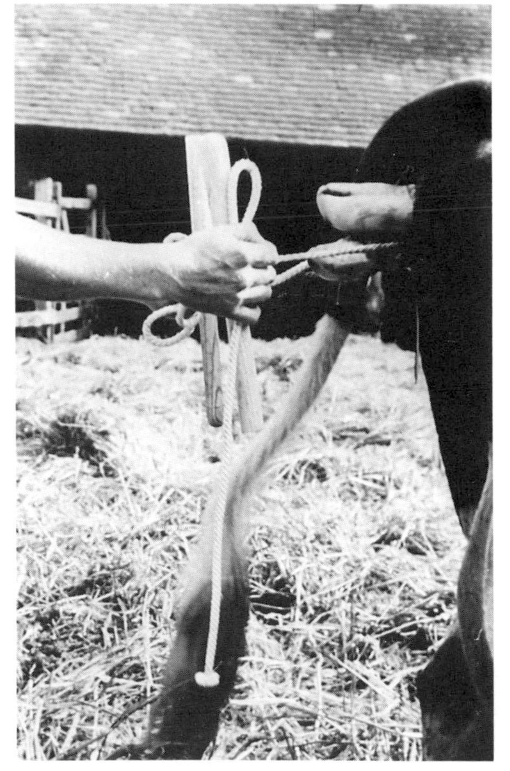

(Above) Calving proceeding well.

(Left) Calving with assistance, using clean sterile ropes.

down, which then involves an examination to feel for the hocks and tail. In this position, once the hips are through, a firm steady pull is required as delay with this form of presentation will lead to a suffocated calf.

Head out and legs back is a complication which requires veterinary assistance, as does breech presentation.

Care of the Newborn Calf

Following birth, it is important to ensure that the calf starts to breathe; otherwise smart action should be taken. Several possibilities are available:

- Push a firm piece of straw into the nose.
- Briskly rub the calf with a handful of straw.
- Slap briskly on the side of the rib cage.
- Throw a bucket of cold water over the calf's head.
- Artificial respiration—laying the calf on its side and working the upper leg round in a circular motion.
- Kiss of life—blocking mouth and one nostril, blowing into the other whilst alternately pressing on the ribs to deflate the lungs.
- If the birth has been protracted and there has been any possibility of placental fluid being inhaled, the calf should be lifted by the back legs and swung around to drain such fluids.

To prevent infection through the navel it is necessary to dip it in iodine solution or to spray with an antiseptic aerosol after licking by the mother is completed.

The afterbirth is normally expelled some 12–24 hours after the calf, its retention leading to infection in the genital tract. If it is not expelled after three days, veterinary attention should be given. On no account should herdsmen pull on the membranes as they may break and the portion remaining in the uterus will then cause inflammation and serious infection.

In some herds there is a high incidence of retained placenta, whereas in others it is a rare occurrence. Abortions and premature calvings can be a cause, as can the birth of twins, milk fever, dirty calving areas or too much manual interference at calving.

Early Lactation

It was stated above that the birth of the calf was a critical time for the cow. That is certainly true, but in terms of the cow settling into lactation, achieving a high peak and total yield, the few days after calving are equally, if not more, critical. As calving is a natural process it should ideally involve minimal interference from the herdsman, but his role really does become important in the days that follow calving in terms of controlling feeding, milking and general management. Avoiding stress in general is a high priority which involves the provision of a clean dry bed, not allowing her to spend more than 24 hours with the calf and moving her into a relatively small group of animals. (It may be worthwhile forming a special fresh-calved group.) One of the main aims at this time is to allow the cow to attain her potential appetite as quickly as possible, so that control of feeding and the avoidance of any digestive upsets are vital. She will require sufficient time to obtain her food and there should be no physical restriction to eating, such as inadequate pasture or lack of trough space. It is important to keep the rumen working well so that a supply of some 6 kg DM from good-quality forage is required daily. Ideally this forage will be identical to that she was receiving pre-calving. Two kilogrammes per day of sugar beet pulp or brewer's grains, because of their high palatability and the relatively slow rate of carbohydrate digestion, will also be a valuable part of the diet. The feeding of concentrates should be very carefully controlled, starting with 3 kg per day and building up over 10–14 days to a level of some 12 kg. Such detailed control should be possible in a cowshed or system where yokes or individually programmed out-of-parlour feeds are available. It is much more difficult to control rations if the freshly calved cows have to join the main herd or a high-yield group where high levels of concentrates are being fed. Few problems should arise in complete diet fed situations as, due to the lower appetite of fresh-calvers, they will not be able to over-eat at this time. When concentrates are fed in the parlour, especially at flat-rate levels, it is vital to identify these freshly calved animals at each milking and to reduce their feed of concentrates to the required level.

Once they are settled into lactation it is essential to meet their energy needs and to offset their continuing low appetite by offering a diet with a high proportion of concentrates. However, to avoid digestive upsets and the possibility of low butter fats, the concentrate: forage DM ratio should be in the region of 3:1 and should not be over 4:1.

Body condition score will drop by 1 to 1½ scores in the first 6–8 weeks of lactation associated with a loss of some 50 kg liveweight. Higher yielders have the ability to eat more food, but not sufficient to meet their increased needs, so they tend to lose more weight than lower yielders.

The important point for herdsmen (and owners) to note is that if this level of care and attention to detail is to be given to the freshly calved animals, and particularly to heifers, then adequate time has to be made available. Lower priority jobs will have to wait or ideally many of them will have been done before the calving season is in full swing. In herds which have a tight calving pattern, the provision of extra help will be well justified at this critical time—as is often provided for a shepherd at lambing time.

The management of the cow during the breeding season will be dealt with in Chapter 11 covering such aspects as oestrus detection and optimum condition score. The role of the herdsman during this time is again a major one in ensuring that the diet continues to be adequate and that the cow suffers minimal stress.

Mid lactation

After conception is confirmed, few problems should arise in managing the cow through mid lactation to regain weight without falling in yield more than 2½% per week. Appetite is now at its peak so that the full potential of background feeds should be exploited. The body condition of autumn calvers will need to be watched at pasture, especially during dry periods, and buffer feeds provided as required.

CHAPTER 6
MILKING

Milking is another major part of the herdsman's job, and his skills and understanding of cow behaviour in performing the job have a considerable influence on the overall success of a dairy enterprise. Cows which are handled quietly and are content at milking time let down their milk effectively, whereas if they are stressed their yields will be reduced and their full potential of breeding and feeding fails to be achieved.

Milking is undertaken by a wide range of people, from herd owners who rarely miss a milking and know every animal in great detail, to the occasional relief who is unfamiliar with the cows as well as the milking plant. In many situations one person undertakes the milking alone, whereas in other large installations two or even more milkers are involved, working the routine together. Many types of milking facility exist, ranging from cowsheds to various types of parlours including a few sophisticated rotary installations. The detailed routine varies somewhat from one type of parlour to another, but since the herringbone parlour is the most popular it is the routine for this type that will be described later in the chapter—but first, a word about the plant that is involved and some other aspects of stockmanship.

Milking Plant

A well-designed and efficiently installed plant is essential to good milking; regular and effective maintenance is equally important. Some of this maintenance, such as replacing worn rubbers, is part of the herdsman's job, but people with specialised equipment need to be called in to undertake mechanical tests.

Some herdsmen work with fairly basic installations of considerable age, which, if well maintained, do the job satisfactorily. Some look with envy at the latest automated plants erected on show stands or on neighbouring farms, whereas others question the potential problems of such sophisticated equipment. Herd owners contemplating the updating of a plant may consider utilising a temporary bail during the alterations, but will need to plan and control such an event in great detail. Cows and herdsmen are likely to suffer considerable stress in such circumstances, so that the changes should be undertaken as quickly as possible and ideally at a time when a high proportion of the herd is dry or in late lactation.

Herdsmen should discuss with herd owners from time to time (but not too frequently!) possible modifications to improve plant efficiency. Only minor expenditure may be required to improve the working environment in the parlour with such things as additional heat in winter or ventilation in summer. Remote control of gates in the dispersal race which divert animals for treatment is an example of a minimal investment which considerably reduces effort and stress on milkers. If increased yields are being achieved or herd size expanded, it may be feasible to incur larger capital sums to increase the potential efficiency or throughput of

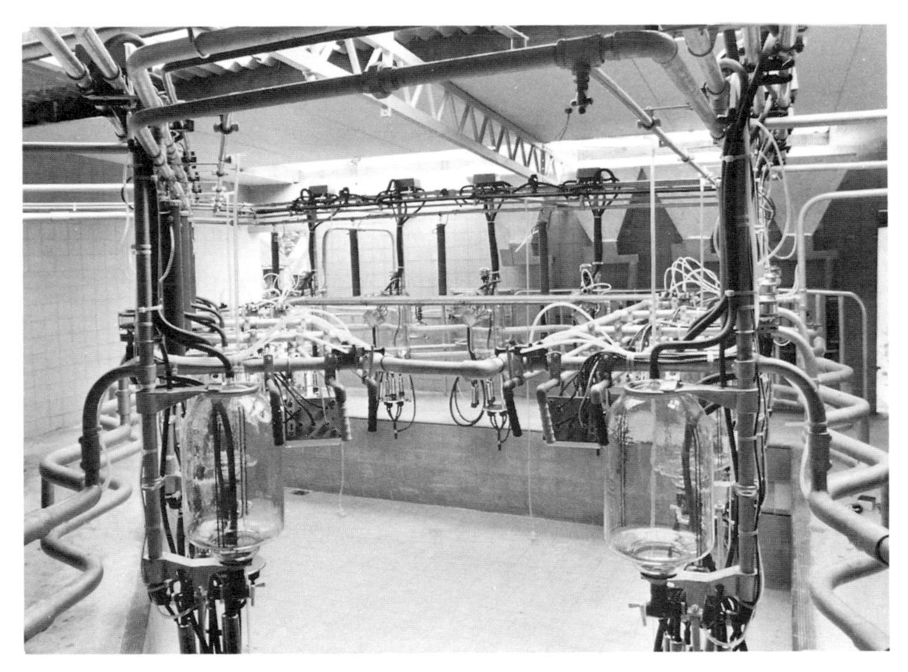

A modern Trigon milking parlour, which is a three-sided parlour with herringbone stalls. This type of layout increases milking throughput by 10 per cent over a normal herringbone with the same number of units. The operator has a shorter length of time to wait for a slower milking cow due to the smaller batch size.

the plant. In a two-stalls-per-unit parlour this can be undertaken by doubling up the milking units or by extending the length to incorporate additional units.

Good herdsmen take pride in their equipment as they do their cows. The building is kept clean and tidy, and there is effective internal and external cleaning of the equipment after each milking session. Such a situation should not be too difficult to achieve on farms where one person undertakes most of the milkings or on larger family farms where the 'boss' does some of the milking. It is perhaps not surprising that it tends to be more difficult to achieve in large herds where a number of milkers are involved, perhaps on a shift system, with divided responsibility for cleanliness. Herd managers in these situations have a key task in setting appropriate standards and ensuring that every operator 'leaves the parlour as he would wish to find it'. Where a team of herdsmen is involved, one member usually has some mechanical interest so that he can, with benefit, be given responsibility for routine maintenance tasks such as the weekly check of oil levels in the vacuum pumps, cleaning of the regulator or adjusting the belt drive on a refrigeration plant.

Milking Staff

It is a basic requirement of satisfactory milking performance that operators are able to concentrate on the job in hand, the milking and observation of the animals. They need, therefore, to have confidence in the equipment, be sure how it operates and know what to do if problems arise. Herd owners and established members of staff should have no difficulty in this area, but relief staff or new employees, especially in one-man operations, often feel vulnerable and do not settle or enjoy the task. Herd owners can help reduce such problems by issuing clear explanations and agreeing a procedure in case of problems. For this kind of reason alone, a telephone in the dairy is a most valuable resource, not only to call for engineers but also for other perhaps more serious emergencies.

In parlours operated by more than one milker the opportunity to train young people is readily available. When two milkers regularly work together, it is a considerable help if their personalities match and they enjoy one another's company. Such problems do not arise in a one-person unit, as he has only himself to blame if he oversleeps by ten minutes one morning; and the dairy radio can always be tuned into the programme of his own choice.

Pre-milking Tasks

There are several important preparatory tasks to be undertaken before the cows are brought into the collecting yard. This will usually apply more to the afternoon than to the morning so that milking can be completed before the transport tanker arrives. Some of these tasks can be undertaken the previous evening if an early start to morning milking is required.

Following milk collection the bulk tank should be cleaned and disinfected. Where automatic washers are fitted the tanker driver will have activated the washing cycle so that only the final rinse may be required. Filters should be assembled, checking for any holes, and the milk supply pipe safely connected to the bulk tank, not forgetting the plug! At this time it is essential to check the temperature of the bulk tank, ensuring that the ice-bank has built up, as it is a contractual requirement to cool supplies down to at least 4.5°C. Where in-line coolers are in use it is necessary to check their effectiveness at each milking by monitoring the temperature of the stored milk. Discolouration of the tank, which tends to build up over the months, can be corrected by spraying (with care and safety) the surfaces with concentrated hypochlorite, followed by thorough rinsing.

It is also advisable to prepare supplies of routine items used during milking, such as paper towels and teat-dip. Fly spray will also be needed on some farms although most now use long-acting compounds or fly-repellent ear-tags. It is also a considerable help to prepare a list of those animals which will need to be diverted after milking for AI or veterinary treatment. A wall-mounted blackboard, visible from the milking pit, is a very useful method of displaying the numbers of such animals.

Parlour and collecting yard floors and walls should be well wetted at this time, so that dung splashes are removed more easily.

The job of bringing the cows from pasture or from winter housing is a most important one in terms of stockmanship. The task should not be rushed, and an adequate length of time should be allowed so that it can be undertaken with care and, therefore, satisfaction.

Detecting cows in oestrus is a key task which can be undertaken at this time since disturbance of the herd often encourages mounting behaviour. When large groups of cows are brought in from pasture, they usually string-out over some distance so that it is difficult to identify a bulling cow towards the front of the herd.

Checking the bulk tank before the commencement of milking.

An in-line milk cooler. This is a very efficient system of cooling large quantities of milk. After cooling the milk, it is stored in a stainless steel tank to await collection.

Enlisting the help of a colleague for a few minutes to meet the herd as it approaches the buildings is useful. Signs of ill health such as lameness or bloat are often more easily spotted at this time. It is also a convenient moment to check water supplies because if a number of cows are seen to be jostling around an empty trough an inadequate supply pipe or trough capacity will need to be corrected.

When bringing the herd from grazing it is necessary to check the amount of forage (grass or other forage crop) which has been eaten and evaluate the supply remaining, especially when a controlled system of grazing such as paddocks or strip-grazing is in use. If an electric fence has to be moved this may be the most convenient time to undertake the job, particularly if the cows can be herded into the track, so as not to interfere with the move. Cows should not be rushed along tracks but allowed to come at their own speed. Concrete surfaces should not be allowed to get too muddy, especially on a farm with a stony soil; otherwise foot troubles will be more of a problem. The lame cow at the back of the herd which delays progress towards the collecting yard should be an encouragement towards more prompt or more frequent attention to feet!

When you bring animals from a cubicle or kennel house you will usually find that a number are lying down and it will take some time for these to get up, back out and move down the passageways. It may be possible to use this time to remove any dung pats from the ends of the cubicles with a fork or even with one's boot, but do not forget to keep an eye open for bulling cows. This is also the best time to check the consistency of the dung as an indicator of optimum diet formulation, and one should always be on the lookout for tell-tale deposits of mucus or 'whites' on the edge of the cubicles from animals which could have disorders of the breeding tract.

Milking Routine

Once the herd is in the collecting yard, milking should proceed without undue delay so as to minimise the time that cows are away from food, water and rest. In some high-performance herds which are milked three times a day, forage as well as a water trough is provided in the collecting yard to reduce this problem.

After switching on the vacuum pump it is necessary to check the gauge to ensure that the working vacuum level is reached

Removing a dung pat from the end of a cubicle bed. This is a job which can often be undertaken whilst the cattle are moving out for milking.

quickly. If there is a delay, check for air leaks, possibly at an open drain tap or a misplaced jar lid. The sound of the pulsators should be regular, and air should be entering the regulator and claw-piece airbleeds.

All should now be ready for letting in the first batch of cows which, in the herringbone, involves opening the backing gate. There is usually some pushing and jostling by those animals most keen to get into the parlour for their feed, when this is given, or to have their milk removed. Some cows are only content when allowed to enter one particular side of the parlour or even one specific place in the batch (usually first or last). Many herdsmen cope with such demands by individual animals, whereas others suffer considerably from such hassle and herd owners would be wise to cull these troublesome 'beasts'.

Where feeding in the parlour is practised, the activation of the dispensing mechanism can usually be undertaken as the cows enter the stalls. In systems fitted with an electronic memory it

Activating the dispensing mechanism for parlour feeding. The number of each cow in the batch has first been entered in stall order. The electronic memory which has earlier been programmed with appropriate feeding levels enables automatic dispensing of each cow's individual ration.

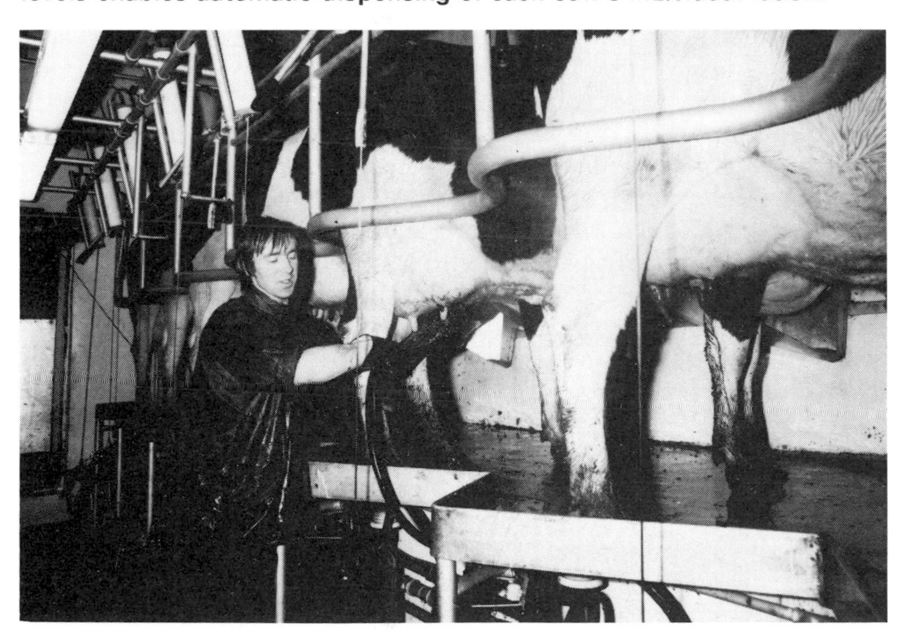

Teat washing from a warm water jet.

is only necessary to press the appropriate cow number for each stall, whereas in other more basic types, after identifying the cow, the herdsman may have to refer to a chart before activating the feeders. The use of coloured tapes fitted to tails or numbered ankle straps can be a beneficial aid to operating a differential feeding system. In large herds subdivided into groups by yield level, or in herds operating a batch calving system, feeding is usually simplified because all cows in the batch receive the same level of concentrates.

Now, by working in a set pattern, that is, starting with either the back or front cow of the batch, the milking routine begins. It is usually worthwhile identifying a slow milker within the batch and dealing with her first, so as not to delay the dispersal of the rest of the batch when they are completed milking. The first operation is to take foremilk from each teat, ideally with a rubber-gloved hand, either into a strip-cup or on to the 'clean' floor. Any animal showing signs of clinical mastitis with clots in the foremilk or one with any milk abnormalities should be marked on both rear flanks with aerosol spray and receive separate treatment, which will be outlined below.

Teats and any soiled parts of the udder are now washed, ideally with a warm water jet, although it should be stated that in many herds where udders and teats can be kept clean and the total bacterial count (TBC) is low, washing has been replaced by dry wiping with paper towel. Handling the udder and teats stimulates milk let-down due to the release of a hormone into the bloodstream. After the teats have been prepared, the teat-cups should be attached. Many animals have a short let-down period so that a delay of teat-cup application will reduce yield. Starting with the back teats, teat-cups are attached by holding the cluster in one hand and using the other to lift each teat-cup in turn, guiding the teat into the cup with one's finger ends. Once the vacuum holds the cups in position, milk begins to flow and one can then move to the next cow in line, repeating the same procedure. In two-stalls-per-unit parlours the opposite batch is prepared while the first one is being milked, whereas in one-stall-per-unit parlours it is possible to let two batches in together and work around the parlour as though with one big group. It is essential to keep an eye and ear open for a cup or even a cluster which may fall or be kicked off the udder.

In parlours fitted with semi-automatic cluster removers it will be necessary to return to each stall after the milk flow is in full stream to switch on the take-off mechanism. Where fully automatic cluster removers (ACR) are in use, it should not be necessary to return to a cow once the teat-cups are applied. Equipment which automatically transfers milk from the holding jars has a short delay built into the sequence which allows the milker time to check milk quality and quantity visually, in case transfer needs to be postponed. It is necessary to observe that cows are milked out properly and if in doubt to feel the udder and draw the teats. In most circumstances the milker will know the expected level of yield and will note those animals which are 'down' at that milking.

In some parlours milk flows directly into the pipeline and no jars are fitted. With such installations, milkers need to be most careful when checking foremilk and, unless milk meters are in regular use, no check on yield is possible.

Where ACRs are not fitted, clusters need to be removed as soon

A herringbone parlour fitted with automatic cluster removers. Once the teat-cups have been applied it should not be necessary for the operator to return to a cow. Note the in-line detectors for mastitis clots in the long milk tubes by the legs of the cows on the left of the picture.

as the milk flow has finished. This involves detailed observation and frequent movement along the pit to deal with individual cows ready for cluster removal. The process involves closing off the vacuum, which prevents any dirt being sucked into the cups, then releasing the milk from the jars—if fitted. Teat cups should not be left on too long, as overmilking can damage udder tissue and also encourages slow milking. If a cow is three-quartered (i.e. blind or dry in one quarter) it is necessary when attaching the cluster to double back the teat-cup or fit a specially designed plug which cuts off the suction.

After teat-cup removal all cows should have their teats dipped or sprayed with disinfectant before being released. A number of highly automated dairies have a set of spray jets installed in the floor of the exit race which are activated by electronic eye as each cow passes through the gate. The milking routine is completed by opening the front gate, usually by remote control, which releases the batch. The opportunity should then be taken to wash away any dung from the standings, a bucket of water being more effective than a hose.

Quiet and confident handling of the animals should usually prevent too much kicking, but problems often arise with newly calved heifers and occasionally with an animal with a sore teat. If help is available the problem can usually be overcome by holding up the tail into an almost vertical position, while at the same time scratching the rump. Tying a rope around the middle of the animal, just in front of the udder, is an alternative. Many herdsmen now use a specially manufactured bar which fits with one spring-loaded end over the backbone and the other under the flank. Milkers should remember to remove the device before the cow leaves the parlour.

The potential throughput of a parlour varies according to the extent to which mechanical and automatic devices are used. In Table 6.1 a full manual routine for a herringbone of 1.2 mins/cow gives a potential throughput of 50 cow/manhour. In each subsequent column one such aid is added to show the effect on the work routine and milking performance. Automatic cow exit can be obtained by linking the exit gate to the removal of the last cluster. As yet no automatic cow entry is in commercial production.

Footbaths

The regular use of a footbath is helpful on farms where foot problems are common due to such factors as soft feet in winter,

Table 6.1 Effect of progressive mechanisation/automation on milking performance

Elements	*Minutes per cow*						
Let out cow	0.20	0.10	0.10	0.10	0.10	Auto	Auto
Let in and feed	0.25	0.15	0.05	0.05	0.05	0.05	Auto
Foremilk	0.10	0.10	0.10	0.10	0.10	0.10	0.10
Wash and dry udder	0.20	0.20	0.20	0.20	0.20	0.20	0.20
Attach cluster	0.20	0.20	0.20	0.20	0.20	0.20	0.20
Remove cluster	0.10	0.10	0.10	Auto	Auto	Auto	Auto
Disinfect teats	0.10	0.10	0.10	0.10	Auto	Auto	Auto
Miscellaneous	0.05	0.05	0.05	0.05	0.05	0.05	0.05
TOTAL	1.20	1.00	0.90	0.80	0.70	0.60	0.55
Cows per manhour	50	60	65	75	85	100	110
Mechanised barriers/gates							
Semi-automatic feed dispensers							
Automatic cluster removers							
Automatic teat disinfection							
Automatic cow exit							
Automatic cow entry							

Source: MAFF.

or soil type, long walking distances or unsuitable road/track surfaces in summer. Allowing the animals to walk through the bath twice a week as they disperse from milking is ideal, if the facility can be so located. A 5% solution of formalin is commonly used but the new COSHH* regulations will need to be checked and may necessitate a change to the less hazardous copper sulphate solution, usually at 2.5%. It is preferable to construct the bath in two sections, subdivided by an area of raised concrete. The first bath has water to clean the feet before the animals walk into the second 'treatment' part.

* An approved Code of Practice for Control of Substances Hazardous to Health as handled on each individual farm. The scheme is administered by the Health and Safety Executive (HSE).

Post-milking Tasks

Where the arrangements permit, it is preferable to allow the herd to move directly back to pasture or to housing as they leave the parlour. If a public road has to be crossed or slurry scraped from buildings while the cows are out for milking it will be necessary to hold the animals in a dispersal area. The time spent here should be kept to a minimum, so that you may need to move cows before starting to wash down.

The equipment, stalls and building should then be washed. The first task is to brush the outsides of the clusters and jetters in a hot detergent–disinfectant solution. Worn or perished liners and rubber tubes should be looked for and replacement undertaken before the next milking. Where the acidified boiling water (ABW) system of equipment cleaning is in use, ensure that at least 15 litres per unit of water at not less than 96°C is available. (The majority of water heaters are fitted with time-switches.) If circulation cleaning is used the water temperature needs to be 85°C, using the manufacturer's recommendations regarding the quantity of detergent–disinfectant. In some well-designed units, low TBCs can be obtained by using hot water only once a day, with cold water and an appropriate chemical at the second wash. Stall work may need scrubbing to remove dung splashes. A high-pressure hose is well suited for walls, but for floors, a high-volume, low-pressure supply is ideal. At least once each week air pipelines and the interceptor need to be washed and sanitised, and in areas of hard water the pipework must be descaled.

Mastitis

Any cow with a clinical infection of the udder should be milked out with care, ensuring that the milk is not transferred to the bulk tank. Some systems allow the jars to be emptied separately and in others mastitic cows are milked by diverting the supply from the jar into a milking bucket for use in calf feeding. The infected quarter should be stripped out by hand before antibiotic treatment is undertaken. Hands need to be disinfected before dealing with another cow.

It will often be practical to proceed with immediate treatment in a herd where consistent results can be expected from the 'proven' type of antibiotic in use. But where mastitis cases are rare or there are varying types of infection, a 10 ml sample of the foremilk should be taken into a sterile bottle before treatment and sent without delay to a laboratory. The results will indicate what form of treatment is most likely to be effective, whether the infection is streptococci, staphylococci, coliform or mycoplasmas. A full course of treatment must be carried out and detailed records kept as an aid to culling decisions. The wall-mounted blackboard, as described above, is ideal for recording cows undergoing treatment. Data from this board is copied into the permanent records as and when convenient (ideally before going off duty).

The use of long-acting antibiotics during the dry period, as described in Chapter 5, is another recommended procedure in mastitis control. At the last milking it is necessary to clean the teat end thoroughly with white spirit after milking, before infusing the antibiotic. The teats are then dipped in iodophor solution and the udder left to stock up.

A number of other husbandry practices to minimise udder infection are worthy of note. These include letting the cows back from milking in to the feed area rather than to the cubicles or bedded yard to allow time for the teat spray/dip to dry and block the teat sphincter before the animal lies in the bedding. Some stockmen have obtained benefit from using a small quantity of slaked lime at the rear of cubicle beds two or three times a week as a drying agent and disinfectant.

Summer Mastitis

This is a serious infection first seen as a hardening of a quarter or stiffening of a teat and loss of appetite. It is thought to be mainly fly-borne and is usually found in dry cows or heifers in tree-lined, low-lying pastures. Careful inspection of dry cows at least once a day is essential, and if infection is suspected the vet must be called for speedy treatment to avoid loss of the quarter or even death of the animal.

The hallmarks of good stockmanship are an attention to detail where disease or infection is concerned, coupled with patience and care for the animals in all stages of their daily round and a desire to see the appropriate plant operating efficiently at all times, and they are nowhere more important than in the milking routine.

CHAPTER 7
HOUSING

A range of well-designed buildings can be of considerable assistance to the herdsman in the efficient management of his cattle. Providing such a facility anew normally represents a major capital investment, so to justify it, enterprise output needs to be increased or costs reduced, or both achieved. Direct benefits can be attributed to housing, as for example the improved growth rate and feed efficiency of beef cattle which are yarded rather than outwintered. More often, the benefits are indirect, due to herdsmen being able to carry out their duties in a better environment. It is possible, for example, to obtain a low Total Bacterial Count of milk from an outwintered dairy herd but this is much easier to achieve if the cattle are housed in well-managed facilities.

Cattle can in fact cope remarkably well outdoors with both cold and wet conditions, especially if they are well fed and can find some shelter from the wind. Food wastage, however, is common in outwintered systems and there is also the problem of damage to the soil from poaching by cattle and the passage of feed vehicles.

Not all farm buildings provide an optimum environment for cattle, especially if overcrowding takes place. Bad ventilation contributes to lung disease, wet bedding to mastitis and slippery floors to split pelvic bones. The layout of the buildings can also have a marked effect on the efficiency of operation. The modern unit has not only regular movements of cattle to provide for but also for considerable vehicular traffic. A typical problem is the turning of tractors and forage boxes at the ends of buildings with an inadequate width of concrete apron. In such situations damage is often caused to buildings and equipment as well as considerable frustration to drivers.

The majority of herdsmen have to work with an established set of buildings and little opportunity, at least in the short term, to influence major changes. They can, however, by using their skills of stockmanship make the best use of available resources to enhance animal comfort, health and performance. If and when building alterations or new facilities are being planned, the herdsman and herd owner should refer to specialised texts for detailed information on design and layout. This chapter is written particularly to help the herdsman utilise existing buildings in his day-to-day work. Herdsmen should nonetheless have much to contribute to the design team of new or updated facilities. They will be able to comment on the factors which will influence their performance as well as that of the animals. Typical requests will be for simple, reliable systems (e.g. slurry handling) and for attention to detail by builders in such areas as the finished surfaces of concrete floors.

An efficient lighting system is a facility which can be of great benefit to stockmen as well as to the stock. Late night observations and the movement of stock, as well as early morning winter feeding, can then be undertaken much more effectively. Cows are obviously assisted in moving to and from feed areas if a light is left on. Some research has even indicated that milk yields have increased following improved lighting, no doubt by appetite stimulation.

An effective handling facility is also an essential requirement of any cattle unit. A race leading to a crush is required for floor inspections (see Chapter 8) and probably for AI (Chapter 11). A crush is also essential for routine procedures such as tuberculin testing, blood sampling and vaccinations.

Maintenance

On some large farms full-time maintenance staff are employed, so that herdsmen have a minor involvement in repairs and upkeep of buildings. On the majority of farms, however, as and when time permits, many herdsmen enjoy involvement in this type of work. This is to be encouraged, so long as the animals are not neglected and the work is not of a highly specialised nature such as electrical wiring.

There is much relevance in the saying 'A stitch in time saves nine' because a broken cubicle rail can easily cause an injury requiring real veterinary stitches. As with machinery, building repairs can only be undertaken satisfactorily if an appropriate set of tools is available. Hammers, saws and some joinery tools will also be required for fencing work, so will usually be on hand. Other more specialist items for such tasks as plumbing or painting will need to be made available as and when required.

As with all a herdsman's jobs, it will help if he prepares a plan for the routine building maintenance. It will be necessary to avoid undertaking such jobs at important times like silage making, straw reception and, of course, calving. Much maintenance work will be undertaken in the summer months when the buildings are not in use. The herdsmen on many farms find it helpful to walk around the farmyard with the owner or manager to discuss possible maintenance and improvements. A person who is not regularly involved in the routine work may notice more easily that facilities such as gate hinges and fasteners are deteriorating.

Before leaving the subject of maintenance and repairs, a few words of advice to employers of herdsmen—especially to those with one-man units. At all times, the number-one priority for herdsmen is the care of the cattle. There may be some times of the year when they have, or could make, time for maintenance jobs; this is good. There will, however, be other times when urgent problems with the herd arise in addition to the routine husbandry tasks. These may be accompanied by additional work on the buildings such as burst water pipes in winter blizzards. On such occasions, management back-up in the form of additional help really does matter. Not only is the herdsman pleased to have the help but also to find to his satisfaction that the boss really does care about him and the cattle.

The remainder of this chapter deals with the herdsman's tasks which he undertakes in the various types of buildings.

Cowsheds

The cowshed, byre or shippon is still widely used on many smaller farms. Dairy cattle are milked and fed in the building as well as being tied in the stalls during the winter months. On most farms, cowsheds are old buildings of substantial construction which are difficult and expensive to modify or expand. Pipeline transfer of milk to the dairy has widely replaced the carrying of buckets but much hard work is still involved in feeding and manure handling. Some cowsheds which have a double row of stalls and a central passage can be cleaned out with a tractor and scraper. Others have automatic manure scrapers fitted in the dung channel, moving material from the building and loading directly into a manure spreader. In each case, serious thought to the economic handling of materials can pay good dividends.

In order to produce clean milk the floor and stalls need to be frequently washed. The beds, however, need to be kept dry,

Cowshed housing.

which involves the herdsman in regular attention to the litter. Oestrus detection is more difficult than with loose housing so that it may be necessary to turn the animals out into an exercise yard to allow mounting to take place. The cowshed is a good environment in which to train young herdsmen because it allows easy identification, individual feeding and close contact with the animals.

Strawyards

This can be a satisfactory method of housing cattle, especially in areas where straw is readily available and where farmyard manure is of value in an arable cropping system. It will no doubt become even more common following the banning of straw burning and possible restrictions on slurry spreading. It is an example of a facility which the herdsman has to use to the best of his ability. He may be able to overcome some design faults such as a large open side, without major expense, by fitting plastic netting. A large building can sometimes be subdivided to reduce competition between the animals, save straw and improve performance.

Stocking density in straw yards is a critical factor which should be under the control of the herdsman. Where 'surplus' animals cannot be found alternative accommodation, some culling may be necessary or perhaps contract rearing arranged if the 'spares' are replacement heifers. A useful guide to stocking density for dairy cows is to allow six square metres per animal. The actual optimum will depend upon size of cow, quality of straw and particularly on the shape of the yard. Long, narrow layouts are preferable to square yards, as there is less cow traffic between bed and feeding area. Ideally the feeding fence is separated from the bed by a concrete strip which can be scraped at intervals.

Straw requirements for bedding as well as for any feeding purposes need to be budgeted in good time. Early-combined straw is usually of higher quality than that harvested later, so the storage site should be prepared and staff be ready to receive and stack supplies. On dairy or beef farms which also produce cereals it is to the herdsman's benefit to cooperate as fully as possible with the arable staff, possibly helping with transport or bale loading.

Ideally, traditional-sized bales are stored alongside the yard where they are to be used, avoiding the need for a vehicle when bedding-up. For outside stacks it is worthwhile ensuring effective covering to minimise the number of wet and broken bales to be handled in the busy winter months.

When a tractor and trailer have to be used during the bedding-up operation, this should be a two-man job if possible, especially if it has to be undertaken with cattle in the yards. No herdsman should need to be told of the need to keep a sharp pen-knife to hand, but perhaps more need to be reminded to have a convenient receptacle for used strings. Strings on the ground are not only unsightly but also dangerous to man and beast and a considerable menace if they are allowed to become entangled in machinery. Where large bales are in use it is necessary for the handling vehicle to be available at times convenient to herdsmen and stock. This is another example of the need for planning and for good team spirit when a vehicle, widely used for general farm handling duties, is also essential to the herdsmen in the bedding-up of yards.

In most situations it will be necessary to clean out the yards at least once during the winter housing period. This is yet another task justifying careful planning and execution, so as to avoid damage to buildings and fittings or undue disturbance to stock. Water troughs at this time are liable to be damaged unless they are well located, and they will also need cleaning out after the operation. This is a job which is often neglected, but should

A large, well-designed cubicle house. Good ventilation is provided by the slatted boarding along all four walls of the building. Note the adjustable panels over the gates at the gable end.

not be: cattle do prefer to drink from clean troughs. The task is made easier if, before cleaning, the ball-cock is tied up for a period so that the animals reduce the amount of water to be baled out.

Kennels and Cubicles

This is considered by the author to be the most satisfactory method of housing milking cows. It can also be well suited to heifer rearing—but it is not as suitable as straw yards for beef cattle or down-calvers. Although lighter construction makes the capital cost of kennel housing much less than that of general-purpose yards (with straw bedding or with cubicles), maintenance costs are much higher.

The ideal kennel or cubicle building should have no more than 30 cubicles on each side of the central dunging passage. Slurry removal from longer 'runs' is inefficient and unless several cross-passages are available between feed and sleep, much bullying can take place. Cubicle length and width need to match cow size and the dunging passage should ideally be 2.40 m wide. For large Friesian/Holstein animals the recommended size of cubicle is 2.20 m long and 1.20 m wide. There is again a limit as to what the herdsman can do to overcome unsatisfactory design but there is considerable scope for efficient management, especially of the beds. Well-cared-for beds improve cow comfort, cleanliness of udders and minimise the evidence of injuries. Where head rails are fitted, accurate adjustment reduces the amount of dung falling on the beds and leads to less work as well as to healthier udders. The base of cubicle beds needs to be firm, with no holes allowed to form, thus reducing the risk of cows being trapped under the bottom rail of the divisions. Rammed chalk, when available, has proved a suitable base material but a thin (75 mm) covering of concrete makes a suitable alternative.

New supplies of bedding need to be provided several times each week. The type of bedding material used on a particular farm will depend upon such factors as location (availability and price) and method of handling slurry. Straw and sand are both good bedding materials except where slurry is pumped. Machines are now available which finely chop straw from the bale (even big bales) and conveniently blow the material onto the beds; slurry can then be pumped. Sawdust or wood shavings are a good buy in some areas, although coliform mastitis has been associated with the material from some types of wood. Rubber and other specialist types of mats are used by a few farmers as a method of reducing the need for bedding material. However, in addition to having a high capital cost, some mats tend to provide a hard and therefore relatively uncomfortable bed and if allowed to retain moisture (under or above) create an ideal micro-climate for bacterial growth.

Dunging passages need to be scraped at least once a day to prevent animals having to spend too much time standing in slurry. In the majority of situations a tractor and rear-mounted scraper is used for this purpose—fitted with a replaceable rubber blade. Some farmers, keen to replace labour with capital and at the same time wanting to minimise disturbance to the cattle, utilise automatic scrapers. Many such units are activated several times each day by a time switch. A safety device is built into the system in case the scraper blade meets an obstruction such as an incumbent cow. Newborn calves do not inactivate the mechanism, so that down-calvers need to be moved out in good time. The author is very much against down-calvers being in any

Grooved concrete to prevent the floor becoming slippery after years of constant use—particularly of the slurry scraper.

Comfort-stall cubicle divisions.

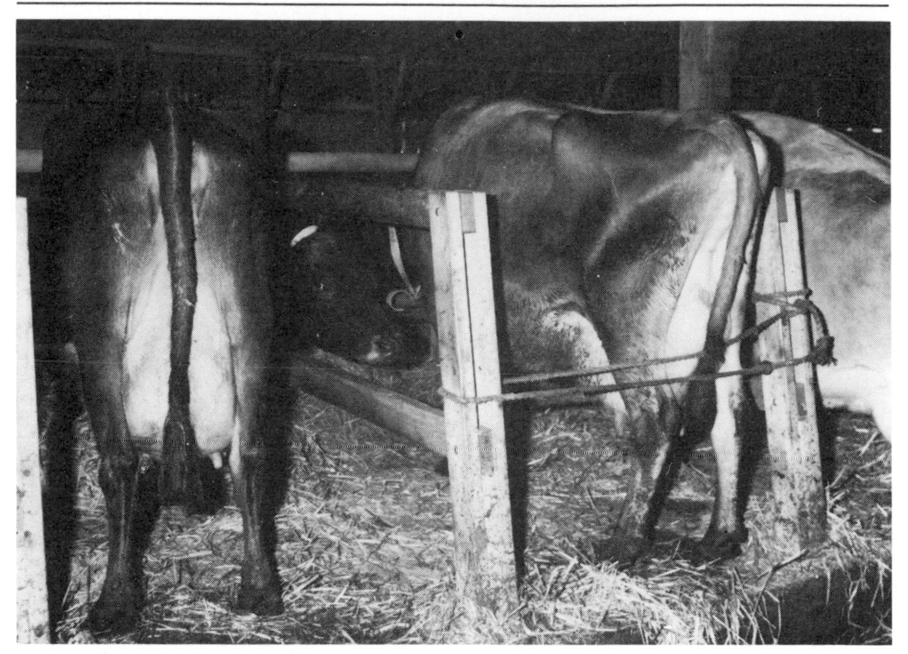

Training a cubicle-refuser. Such problem animals are rare if cubicle housing is used in the heifer-rearing stage.

type of cubicle building, having been involved in retrieving newly born calves out of the slurry!

Returning to the subject of tractor scraping, it is rather an unpleasant job on many farms, especially where there are poor layouts, narrow doorways or uneven concrete surfaces. It is a job which has to be done day in and day out, so that care and attention need to be given to both tractor and scraper (see Chapter 13).

Following continuous use and especially scraping over a number of years, some concrete surfaces become smooth and slippery and then are a hazard to man and beast. It is possible to have such surfaces grooved by a special machine (usually a contractor's job). To avoid the problem arising, during the preparation of a new concrete floor it is advantageous to insert grooves in a diamond or rectangular pattern.

Scraping should ideally take place when the animals are out of the building but they should not be moved solely for the scraping process. Considerable organisation is necessary, especially in larger herds, to ensure that the job is done at times convenient to cattle as well as staff. On very large units it is almost a full-time job, at least for major parts of the day. A problem can arise if the task is regularly delegated to a junior herdsman or student, who will easily lose motivation unless given some additional responsibility.

When introduced to cubicles the majority of cattle soon take to the system—but a few do not and present a problem, or perhaps a challenge, to herdsmen. Frequently these cows lie in the dunging passage so that on arrival in the milking parlour they are a genuine problem. It is worthwhile encouraging them to use the cubicles by offering hay in the front of the beds. Some herdsmen claim success from tying offenders into the cubicles for several hours. Others pen the cubicle-refusers to one end of a passage and place concrete blocks in the slurry so that the animals cannot lie down. Persistent offenders are better sold to farms which use non-cubicle housing. It is a fact that heifers which spend some of the rearing period in cubicles are less prone to this problem.

Calving Areas

Cows calve in a wide range of situations from pasture to specialised maternity pens. If a cow is to calve in a cowshed it is wise to tie the animal's neck chain or rope to the stall with a thin piece of string so that in case she is down and slips back, the string will break and she will not be choked. Traditional calving boxes have a considerable labour input so that on many farms calving takes place in a covered yard. The calving of healthy cows on a straw bed creates few problems and at least the cow is unlikely to slip and split her pelvic bones. Facilities to subdivide the area are essential for pedigree herds and worthwhile in other situations, so as to ensure accurate matching of offspring to dam.

An ideal layout is a narrow building, subdivided into pens, with gates which can be folded back for mechanical manure removal. Pen fronts are also formed by gates which carry a feed trough and water bowl. When opened both allow the cow to enter the pen and also block the remainder of the passage so that a cow can be moved single-handed. If the passageway in front of the pens is kept tidy, cows can be checked out of working hours in low shoes, perhaps on the way back from an evening out.

TV cameras are in use on a few farms to save the herdsman time and effort in checking cows due to calve. Pictures from the pen are transmitted to a small screen in the lounge or bedroom.

Maintaining a clean, dry bed is essential in calving areas to minimise disease. At the first sign of infection it is essential to move the animals out and thoroughly clean and disinfect the area. During the non-calving months, it is advisable to leave the facilities clean, disinfected and free of animals.

From all that has been written in this chapter it can be seen that farm buildings can be a very useful aid to herdsmen in carrying out their routine duties. Care and attention need to be given not only to the upkeep of the facilities but also to ensuring an optimum environment for the stock.

CHAPTER 8

HERD HEALTH

Herd health is an area in which the stockman has a major influence upon the performance and productivity of the enterprise. Ideally all the animals in a herd would be free from disease but it is clearly accepted that such a situation is not likely. A survey carried out by the Ministry of Agriculture indicated that over 40% of the cattle in the national dairy herd of Great Britain are at any one time affected by disease. The stockman, therefore, has a vital role in observing the stock, identifying problems and, most of all, nursing the sick back to health by carefully implementing a control programme.

The diseases are seldom due to a single cause but rather to two or more factors, including housing conditions and the quality of nutrition, both of which are considerably influenced by stockmanship. Some health problems, together with their effects on performance, are obvious (for instance a lame cow). On the other hand, many others such as a sub-clinical mastitis can remain undetected for long periods of time.

The role of veterinary surgeons in the detection and control of disease is well recognised, and increasingly they are involving themselves with farmers and herdsmen in a team effort to promote herd health.

To achieve success with such programmes the team needs to meet at intervals to plan and assess performance. They should arrange for:

- Implementation of a set of management practices which reduce the risk of disease, e.g. dry cow therapy or anthelmintic dosing programme.
- A recording system which pinpoints problem areas such as conception rates or youngstock weights.
- Regular veterinary visits especially at critical times such as the breeding season, e.g. pregnancy check and examination of cows not seen in oestrus forty days after calving.

In this chapter we are concerned with the stockman's role in health promotion. It will be a considerable help if he understands the structure and function of the animal body and he is referred to appropriate textbooks for such details. An ideal one is Roger Blowey's *A Veterinary Book for Dairy Farmers*, which is a very comprehensive but readable text, not just on the functions of the cow but on all cattle health topics. The stockman's role is to provide good husbandry so that the natural functioning of the animal is promoted by adequate feeding and water supply, healthy, comfortable surroundings and optimum group size. He needs to have adequate isolation facilities and comfortable areas in which to house sick cows when they are separated from the herd. He should be skilled in detecting the symptoms of ill health, this being achieved by training and experience. The role of older stockmen cannot be overemphasised at this juncture in pointing out to newcomers the detailed characteristics, often rare, of specific health problems. Developments are taking place in training aids such as a dummy cow and calf to provide invaluable experience in 'calving', which is such an important aspect of the herdsman's job. The video medium will no doubt be used in the future to show trainees examples of health problems, and particularly methods of dealing with them, which would otherwise take years to experience.

He must develop a working routine so that the animals have a regular pattern to the day in milking and feeding, fitting in other tasks when possible. Although cows spend a considerable amount of time resting, especially at night, they actually sleep for only a few minutes, rarely stretching the head and neck out flat. It appears that cows need to remain vertical so as to allow cudding to function properly and to prevent bloat.

Signs of Health

At pasture, an animal which wanders away from the herd may be in ill-health, although when operating a set-stocked grazing system the herd does tend to be more scattered so that abnormal posture may be a better way of identifying a sick cow. She may be standing with a dejected appearance—ears and head held low and back arched, or, when lying down, the head either stretched out or turned towards the flank. In general, healthy animals are alert and if they are disturbed their response is reasonably quick. Cattle that are dull and listless should be watched for metabolic upsets, acute mastitis or a foreign body in the stomach wall.

A healthy animal with moist nostrils, bright eyes and a shiny coat. Note the metal device in the nose to dissuade her from suckling other cows—a bad habit developed by a few animals.

The skin of a healthy animal should feel soft and pliable when handled over the last rib; a tight skin and lack of bloom are often associated with feverish conditions or digestive upsets. Healthy animals show lick marks on their skin, and the tail and hind-quarters are free of dung contamination. The dung of a healthy animal should be reasonably formed, depending upon diet and therefore on time of year. Very loose or very firm dung indicates a possible digestive upset. Strong-smelling or blood-stained dung and urine are abnormalities, as are coloured discharges from the genital organs.

The nostrils of a healthy animal are moist and free of mucus and the eyes should be bright. Sunken eyes and a staring look are often signs of the onset of disease.

Holding the ear of an animal is another good health check and it should feel 'warm'. Any fall in milk yield indicates some problem with the metabolic processes and further investigation should take place. Any abnormality in the appearance or smell of the milk should be noted, as the smell of acetone (pear drops), for instance, indicates acetonaemia.

THREE HEALTH CHECKS

There are three particular checks on health which the stockman can use:

1. Temperature

Amongst its many functions, the blood system of an animal regulates the heat of the body, and in a healthy cow the temperature rarely varies more than one degree. The normal temperature of a cow is between 38 and 39°C and is taken by placing a clinical thermometer in the rectum.

The thermometer should first be well shaken to ensure that the mercury is down into the bulb, lubricated before insertion and then held slightly sideways in the rectum so that the bulb makes contact with the mucous membrane. Modern thermometers need only to be in position for some 30 seconds before they can be withdrawn and the reading taken.

Body temperature may be raised by bad housing, exertion of the animal—especially in hot weather—or by fear resulting from handling; but the main cause is fever.

Veterinary assistance should always be obtained if the temperature is over 39°C (103°F).

2. Pulse

This measures the rate at which the heart is beating and is taken by using the index finger of the right hand with gentle pressure on the artery underneath the tail. The normal rate is 45–50 per minute and a rate of over 80 is dangerous.

3. Respiration

This can easily be obtained by counting the number of chest movements occurring in one minute. The rates vary widely; animals breathe faster when young, following exercise, in hot weather and when frightened. The rate for a resting cow is 12–20 per minute.

Metabolic Profile Tests

Several disorders can be prevented or controlled more effectively by having information about the blood components of the cows in the herd.

Blood samples are normally taken from three groups within the herd: dry cattle, high- and low-yielders. Laboratory testing produces a comprehensive report on the levels of major blood components such as glucose, albumen (protein), calcium, phosphorus and magnesium. The interpretation involves comparison with standard figures and is best done in conjunction with the veterinary surgeon taking into account milk yield, body conditions and the composition of the diets being fed.

Several major health problems are discussed elsewhere in this book, such as mastitis in Chapter 6 and infertility in Chapter 11, but a number of the other more common ones will now be covered.

TACKLING COMMON TROUBLES AND DISORDERS ON THE FARM

Bleeding

To stop bleeding, make a pad of cotton wool, lint or even a handkerchief, moisten it in warm water and hold it firmly over the wound until a bandage can be applied to hold the pad.

A wound caused by a sharp instrument has relatively smooth edges and bleeds more profusely than one more jagged (as from barbed wire) which may take longer to heal although the bleeding is less profuse. If the blood tends to spurt out, then an artery has been damaged. Apply a tourniquet and call for immediate veterinary help, keeping the animal as quiet as possible.

Wounds

It is necessary to clip the hair from around the wound before cleaning up the area with a salt solution (two teaspoonsful of salt to one litre of boiled water). Damaged teats are a particular problem and may require stitching by the veterinary surgeon, especially if the teat-canal is penetrated. If wounds cannot conveniently be bandaged to keep them clean, then a dusting with sulphanilamide or antibiotic powder will minimise risk of infection.

Bruises

Bruises, including swollen joints, should be bathed with cold water followed a few hours later by warm water to stimulate blood circulation. If seen early by the veterinary surgeon, good control can usually be obtained. Good results are often obtained by using an Animalintex poultice.

Choking

If an animal has an obstruction in the gullet, caused for example by a potato, it is advisable to try and locate the obstruction by feeling along the length of the underside of the neck on the left side of the wind-pipe. Massage towards the throat will in most cases dislodge the obstruction, but more difficult cases will require veterinary aid. *Never* use a broom handle or other instrument to push the obstruction down.

Downer Cow

Over the years, stockmen will have to deal with, and therefore care for, cows which cannot get up and are said to be recumbent. It is a very rewarding experience to have such a cow get to her feet after careful nursing but equally saddening to see one destroyed following days or even weeks of attention.

The most common cause is milk fever (see below) but others include acute mastitis, severe loss of blood and nerve damage following a difficult calving. If the cow goes down on concrete, e.g. in a cubicle passage, she should if at all possible be moved to a soft bed in a yard, or even at pasture. It is wise, before attempting to move her, to tie the back legs together with rope at the fetlocks or to use a special chain fitted with leather ankle straps. This procedure prevents the cow 'doing the splits' and damaging her pelvis, as she could well have a damaged obturator nerve—which is the one controlling the muscles that hold the back legs together.

A traditional method of moving a downer is to use a gate or

large door dragged by a tractor. This needs to be done with care and adequate assistance to avoid damage to head, udder or lower legs. Special harness is now available which acts as a hammock that can be lifted by a tractor loader.

Veterinary assistance will often be justified but before phoning it is helpful to check the udder for acute infection and also to have the facts to hand, e.g. calving date, number of lactations, to help initial diagnosis.

Good nursing involves regular turning and generally keeping the cow as comfortable as possible with liberal bedding until she is back on her feet.

Eye Problems

Cattle sometimes get small pieces of chaff or other objects in their eyes, causing inflammation and a discharge. It is usually a tricky job to remove the chaff with the corner of a clean handkerchief, but slightly easier with a little eye ointment on the end of a small tube nozzle. If after several attempts the foreign body remains it is better to call the veterinary surgeon who will probably use local anaesthetic to aid removal.

An eye disease of increasing importance in cattle of all ages, but especially of yearlings, is New Forest disease. Infected animals should be segregated, kept in the shade and protected from flies. Treatment with antibiotic ointment or drops may be necessary for several days although with mild cases immediate control often follows. It may be advisable to treat the whole group if several are showing signs of infection. Prevention may soon be possible following the development of effective vaccines.

Bloat

This is a condition caused by gases building up in the cow's rumen due to abnormal fermentation usually caused by lack of fibre in the diet. The left side of the animal becomes distended between the last rib and backbone or hip joint, breathing becomes laboured and prompt action is needed. It most commonly occurs at pasture so that with mild cases the animal can be walked quietly back to the buildings. A drench should then be carefully and slowly given of 28 ml of turpentine in 0.6 litres of boiled linseed oil (usually bought ready mixed) or a proprietary drench. Gentle massaging of the extended rumen may help in the expulsion of gas by belching. If, despite drenching, the bloat is getting worse it will be necessary to puncture the rumen wall. Traditionally a trocar and cannula has been used but a wide-bore syringe needle can be just as effective and leaves a smaller wound to heal. Care needs to be taken to locate the correct position for making the puncture. It is on the left side equidistant from the last rib, the point of the hip and the bottom of the spine (see Figure 10).

Figure 10. The correct site for puncturing the rumen wall.

In emergencies any sharp knife can be used, making sure that it penetrates at a right angle to the skin. It is essential to be bold, and not dither; otherwise a dead animal will be the result.

Prevention should be the key factor in this disorder, which is markedly influenced by stockmanship. Avoid turning animals out on to young lush pastures unless they have received a good feed of fibrous material such as hay or straw. When bloat is experienced with strip-grazing it is a help to move the fence more than once a day so as to prevent the animals getting too hungry and then gorging themselves. The grazing of clover and lucerne is more likely to cause bloat than grasses so that it may be necessary to cut and wilt these crops before grazing in front of an electric fence.

Displaced Abomasum

This is an increasingly recognised ailment which involves the fourth stomach or abomasum moving from its normal site on the lower right-hand side of the abdomen to the left, under the rumen. The specific cause is not known but it is thought that the abomasum is displaced during the first six weeks of lactation. The cow is usually off her food so that the secondary symptoms of acetonaemia can confuse the diagnosis and veterinary advice is required. Surgical treatment has a high success rate and within twelve hours of the operation the cow is usually eating again, if carefully nursed.

Grass Staggers

This is a disorder caused by a deficiency of magnesium in the blood and is also known as hypomagnesaemia. Magnesium helps to control muscle function and an animal with deficiency becomes unnaturally alert and its muscles twitch. It develops a staggering gait and may soon collapse, and convulsions will follow. Death can take place in a very short time unless treatment is given, i.e. a subcutaneous injection of magnesium solution. The disease is most likely to occur in spring and autumn when a sudden drop in temperature occurs and when herbage, especially from young leys, tends to be low in magnesium content. Preventative measures involve the supply of an adequate daily level of the mineral which is conveniently provided in 60 g of calcined magnesite mixed with molasses or rolled cereal or in purchased concentrates at specified inclusion rates.

Milk Fever

This is another metabolic disease caused by a low level of calcium and often magnesium in the blood. Known as hypocalcaemia it usually occurs in older cows (third calves and above) around the time of, or in the few days after, calving. The first symptoms are paddling with the feet followed by loss of control of the limbs. The cow is off her food and ears are cold to the touch, but temperature is normal. An injection with calcium borogluconate under the skin of the neck should be given but it is always advisable to call for veterinary help as there may be additional deficiencies such as phosphorus or magnesium.

When a cow is down for several days she should be kept comfortable by liberal bedding, the brisket propped up and frequently turned to prevent pneumonia and pressure sores. As an aid to lifting a recumbent animal a large cylindrical bag is available on the market which can be placed under the cow and then inflated with a 12 volt battery-powered air pump. Routine injection with calcium after calving may be a wise precautionary measure from the third calving onwards.

Acetonaemia

This is basically a digestive problem caused by the presence of excessive ketone bodies in the blood. These are the by-products resulting from mobilisation of body fat reserves which the cow has to undertake when the diet is low in readily available energy. The first sign of the disease is a drop in milk yield and a loss of appetite, usually for concentrates rather than roughage. The breath and milk of the cow have a characteristic sweet smell of acetone (pear drops). Temperature is normal but the cow is constipated and the dung is often slimy in appearance. Treatment involves the injection of simple sugars together with drenches of laxative or sodium propionate, which helps the production of natural sugars in the rumen. It is helpful to feed carbohydrate-rich materials such as flaked maize.

This is another disease where prevention is particularly important and involves the feeding of a balanced and acceptable diet containing not only sufficient digestible energy and proteins but also adequate fibre. The skilled feeding of concentrates in the early weeks of lactation to avoid digestive upsets but still obtain peak yields should minimise the number of cases of acetonaemia.

Lameness

This is a problem in most herds and a major one in many, so that herdsmen need to be quick at recognising and treating it. There is a saying 'No foot—no cow', which is very true, as, because of the pain, a lame animal is reluctant to move to obtain food and water. In addition to causing loss to the farmer in meat and milk production, lameness also leads to the premature culling of valuable breeding animals.

Lameness is caused by many factors including faulty nutrition, poor buildings and badly laid concrete, as well as by the genetic make-up of the animal. Some 9 out of 10 cases are due to inflammation or injury to the hoof or skin between the hooves. The remaining cases are caused by arthritis, swollen knees, hocks or leg and muscle injuries.

Sound feet and legs should therefore be a factor in the selection of breeding stock but care and attention in the day-to-day management of feet is a key task for the herdsman. When visiting herds as a competition judge, one of the first areas I look at is the condition of feet—entrants take note! Such a warning will not be necessary for good stockmen, as rightly they take pride in this area of their work. Young herdsmen are encouraged to attend specialised training courses as organised by the ATB.

Regular foot trimming needs to be undertaken, as hooves grow continuously, and although some will wear naturally, many need to be trimmed. Professional foot trimmers with wide experience and specialised equipment have become very much part of the dairying scene. However, with a suitable crush and an appropriate set of hand tools, many herdsmen now undertake their own trimming as and when they can find (or make) the time. An ordinary crush is adequate when handling small numbers, using a rope with a slip knot around the hock and pulling the leg up tight to the side bar of the crush. Specialised crushes are available with winding gear which involves putting belts under the cow's belly and lifting her off the ground. These certainly make the task of dealing with front feet easier. Contract trimmers tend to use even more sophisticated equipment, strapping the animal to a rotating table operated from the hydraulic pump of a tractor.

Once the animal is immobilised in whatever crush, the foot needs first to be washed and brushed clean. Using hoof clippers and a special foot knife, ideally of the double-edged type, the overgrown wall and toe of the hoof is trimmed in thin shavings—without drawing blood. The outer claw usually needs more trimming than the inner one, but both need to be left with the weight-bearing surface back to normal.

The same procedure of restraining the animal and cleaning the foot is undertaken when investigating lameness. There are several causes of lesions in the foot:

Foreign bodies A nail or sharp stone can penetrate the hard sole. This can usually be removed with a hoof knife. Infection often follows, but with adequate drainage from the site, improvement should soon follow.

Foul-of-the-foot This is a bacterial infection which enters through cracks in the skin between the two claws, usually causing pus and a putrid smell. The foot becomes hot and swollen and the cow goes lame very quickly. Veterinary treatment involving antibiotic injection is usually necessary.

Sole ulcers These appear as soft bruised areas at the junction between the sole and the heel of the outer claw due to excessive weight on the area which is caused by overgrown solar horn.

Ulcers tend to be associated with laminitis, excessive standing and the feeding of very acidic silage. The hoof needs trimming and the ulcer exposure to the air, and copper sulphate is used for treatment. It may be necessary to 'block' the foot by gluing a wooden sole on to the good claw to protect the damaged one.

White line abscesses These are caused by damage, perhaps by a small stone to the tissue (known as the white line) at the junction of the sole and the wall of the hoof. The tissue becomes infected, causing pus which eventually breaks out to the surface. When trapped the pus causes pressure, pain and severe lameness. Antibiotic aerosol can be used, coupled with a cotton wool pad held in place by adhesive plaster.

Other afflictions These include sandcracks, underrun heel, interdigital growths and dermatitis.

As with all foot problems, prevention is better than cure.

Skin Problems

The skin of cattle occasionally becomes infected so that treatment by the stockman is required. Lice cause unthriftiness, particularly in young cattle. Simple dusting along the neck and back line with louse powder is an effective control. Care needs to be taken when using some compounds, e.g. organo-phosphorus, in confined spaces, a face mask being essential to avoid inhaling the product.

Ringworm occurs most frequently in young calves and yearlings and it is caused by fungal spores picked up from wooden fixtures such as penning or gateposts. Thorough annual cleaning and creosoting of wood is necessary, as well as ensuring that the stock receive adequate nutrition. Drugs which can be used as feed supplements or in aerosol form are of considerable help in treating this problem.

Allergies, photosensitisation and warts are other skin conditions which can suddenly appear and often disappear for equally unexplained reasons.

Worms

The chapter on youngstock rearing will describe the problem of parasitic worms. Dosing of young cattle with anthelmintic is therefore recommended in early July, especially if the animals have been grazing old pasture or a sward grazed by cattle in the previous year. Dosing of adult cows for worms may be justified in some situations, generally immediately after calving.

Infectious Bovine Rhinotracheitis (IBR)

This is a fairly recent disease of UK cattle but it is already having widespread effect. The nose and windpipe are the chief sites of the virus infection, but the disease is seen in several forms, also causing eye lesions and even abortion. It can be controlled by vaccination and in most situations this is a justifiable expense.

Leptospirosis

This is a disease of increasing concern to the industry, not only because it causes abortion and mastitis in cattle, but also because the infection can pass to man—thus making stockmen particularly vulnerable. From the results of blood tests, antibody levels indicate that a high proportion of UK cattle have been exposed to infection; however, the clinical disease is much less common. Cows (and people) are infected by urine splashing into eyes, mouth or cuts in the skin.

Milking cows lose appetite, drop sharply in milk yield and have a raised temperature. The udder becomes 'flabby', as though all quarters have mastitis without the swelling. Antibiotics help recovery. In herds with any possible infection, vaccination is strongly recommended, if only to protect the staff.

The symptoms in man are headaches, high temperature and aching joints, with the occasional case of meningitis and even death. As the disease is relatively new and still rare, most general practitioners have had little or no experience of dealing with it, so it is vital that farming people when reporting symptoms point out that they work with cattle.

AILMENTS CONTROLLED BY LEGISLATION

To conclude this chapter, there are a number of health problems which are in some measure controlled by legislation.

Warble Fly

Widespread control treatment by farmers over recent years has drastically reduced this problem so that 'gadding' cattle due to warble fly are fortunately very rare.

Legislation requires that any cattle found to be infested are reported to the Divisional Veterinary Officer. Movement is then restricted until all cattle on the farm over 12 weeks of age have been treated using a liquid systemic dressing.

Sudden Death

Anthrax, which is a notifiable disease, should always be considered following sudden death. The carcase of such an animal is highly dangerous so that it should not be interfered with or disposed of until veterinary advice is taken. Sudden death may be caused by poisoning or by various bacteria but in all cases veterinary advice should be sought at once. Poisoning can be caused by plants such as yew or hemlock or by chemicals such as sheep dip and lead paint.

'Blackleg' can cause sudden death, particularly in young cattle at pasture, so that vulnerable animals should be vaccinated by the veterinary surgeon before turn-out in the spring.

Foot-and-mouth Disease

If this disease is suspected, it is preferable to notify the police or Divisional Veterinary Officer rather than call one's own veterinary surgeon, who in a positive case would be prevented from working for a considerable period of time. The symptoms are lameness, a sudden drop in milk, high temperature and loss of appetite. Later there is dribbling from the mouth, and blisters form on the tongue, gums, muzzle, teats and feet.

Tuberculosis

This notifiable disease is now very rare in the United Kingdom.

Enzootic Bovine Leukosis

This infectious virus disease is present in a small number of imported cattle. It can be identified by laboratory examination of a blood sample and the Divisional Veterinary Officer should be notified.

Bovine Spongiform Encephalopathy (BSE)

First identified in Britain in 1985, BSE is a disease similar to scrapie in sheep which affects the nervous system, particularly the brain—the symptoms occurring many years after the initial infection. Symptoms normally start with a usually quiet cow becoming nervous, then aggressive. Lack of coordination of the hind legs when walking, especially under stress, is the test used by the veterinary surgeon.

Early epidemiological studies indicated that all the infected cattle had been given feed containing ruminant protein as meat and bone meal—a practice which has been undertaken for decades. The method of processing, however, has changed over the years, and sheep numbers have increased so that cattle have most probably been exposed to more scrapie in their feed.

It is hoped that following withdrawal of such protein sources from cattle diets, the disease will die out, but there is a risk of maternal transmission from cow to calf. A ban has been introduced on certain offal for human consumption to minimise the risk of any infection into the food chain. Trials are underway to investigate any likely spread of the infectious agent and the disease has been made notifiable with full compensation.

Current advice is not to slaughter the offspring of positive cases. It is most important that good records of the parentage of all cattle are kept in case the disease is eventually shown to be transmissible.

It is evident from reading this chapter that there is a very wide range of potential dangers to herd health. Nevertheless, with the help of his veterinary surgeon, the alert, confident and capable stockman minimises these dangers so that they are the exception rather than the rule.

CHAPTER 9

CALF REARING

After birth it is preferable to leave the calf with its dam for 24–36 hours to enable it to suck the first milk, called colostrum. On some farms, for convenience, the calf is left with the cow for up to four days but this practice can create considerable stress to both at separation. Calves which are to be reared on the bucket-feeding system learn to drink much more easily when removed from the cow at the earlier time.

The need for satisfactory suckling in the early hours of life cannot be overemphasised. If the calf fails to suckle unaided, it should be guided to the teats and encouraged to suck. It may be necessary to insert a teat into the mouth of the calf and draw milk. If, following such assistance, the calf still does not suckle, it should be given hand-milked colostrum from a bottle.

Colostrum

Colostrum, or 'beestings' as it is often called, has a high nutritive value. It is particularly rich in protein and is a laxative which helps to remove the dung which has accumulated in the gut before birth. The protective substances contained in colostrum are most important, as they give the calf immunity from the bacteria which are inevitably present on the farm. These are known as immunoglobulins or antibodies and are carried in the fat globules of the colostrum. Research work has shown that the rate of absorption of these immunoglobulins from the digestive tract into the bloodstream peaks some three to six hours after birth. This is the reason why early satisfactory suckling is so advantageous.

Following separation from the cow, whole milk should be fed to the calf twice a day for the first week of life. Surplus colostrum from freshly calved cows can be stored by deep freezing or by allowing it to go sour before subsequent feeding to bucket-reared calves. Some herdsmen consider the handling of sour colostrum to be rather unpleasant but calves reared on it appear to thrive satisfactorily. A simple laboratory test has been developed to indicate the level of antibodies in the blood, using a sample from the calf. Known as the zinc sulphate turbidity test, it is a most useful aid in rearing bought-in calves.

Housing

Following removal from the dam the calf should ideally be placed in a clean, dry, well-bedded individual pen. This should be located in a well-ventilated but draught-free building maintained at a temperature of 8–10°C. The relative humidity of the air should be kept at low levels by avoiding the washing of buckets or using excess water within the actual calf house. In a large calf-rearing enterprise, operating for an extended period throughout the year, it is preferable to have several smaller calf houses rather than one large building. This enables an all-in all-out system to operate with cleaning and disinfection between use, which helps to control disease.

The use of slatted boarding under the bedding of the pens can be a help in reducing straw usage by improving the drainage, but it does increase the time required to wash and disinfect the facilities after use.

Natural ventilation is adequate for the majority of calf buildings, providing that condensation does not take place in winter conditions. Power-operated systems using thermostatically controlled fans are installed in some units in order to achieve a satisfactory environment. Air movements need to be maintained even in cold weather. The ideal system is one which automatically heats the incoming air when the outside temperature drops to a set level.

METHODS OF CALF REARING

1. Single Suckling

This is the most simple system of rearing, as the calf suckles its own mother from birth to weaning at nine to ten months of age. It is expensive, especially when using high-grade land, and is only justified for the production of first-class animals for beef or for valuable pedigree animals. It is, however, well suited to low-output upland holdings or to arable farms where the cows can utilise crop by-products. The cow is usually of a pure beef breed or is beef × dairy with sufficient milk production for the one calf.

Single suckling. The bull calf is nine months old and ready for weaning. It was sired by a Charolais and its dam on the left is a Limousin × Friesian.

When crossed with a beef bull such cows produce high-quality animals. Natural service is the usual system of rebreeding suckler cows, although prostaglandins are being increasingly used to synchronise oestrus so that AI can be more conveniently used.

Magnesium deficiency (grass tetany) is often a problem with the cows and vitamin E deficiency (muscular dystrophy) is common in suckled calves. Occasionally a problem can occur when the calf is unable to consume all the available milk, leading to mastitis.

Careful observation is required by the herdsman, and this is much easier in a small herd rather than one with large numbers of cows and calves running together.

2. Multiple Suckling

This system involves suckling several batches of calves on a cow during the lactation. The number reared will depend upon the milk yield of the cow and the number of weeks each batch is allowed to suckle. Using a cow of a dairy breed and allowing eight to ten weeks suckling of each batch, then 15 to 20 can be reared in one lactation. Although used on a few farms to rear dairy heifer calves, this is most commonly used for rearing purchased calves for a beef enterprise. The ideal cow is a high-yielding dairy type of reduced value due to some conformation defect which causes milking problems in a modern parlour system. Wide teat placement or loss of one quarter is a common reason for culling from the dairy herd so that if the animal has a quiet temperament she would be ideal for rearing calves. It is often difficult to identify oestrus so that the cow fails to re-breed and is culled.

Considerable patience is required by the herdsman when persuading the cow to take to her calves, especially to a new batch following weaning of the previous group. Some hand milking may be necessary for several days at this time until the new calves are able to consume all the available milk. With some cows this can be a rather challenging task and a good test of stockmanship! It is a help if the cow can be tied when mothering-on new calves and it may be necessary to tie a rope around the abdomen in front of the udder, to stop kicking. It is advisable not to include the cow's own calf in the first batch, moving it to another cow once it has obtained colostrum.

This is a system well suited to a small-scale rearing enterprise producing quality weaned calves at periodic times throughout the year. Health problems tend to be similar to those of single suckling, although the incidence of nutritional scouring is often higher.

3. Bucket Rearing

This is the most widely used method of rearing calves, for both beef production and dairy heifer replacements. It is a suitable system for rearing home-produced and purchased calves but in the latter case much more care is required to minimise disease problems. The system is based upon the use of purchased powdered milk replacers which are mixed with the appropriate amount of water and fed from the age of seven to ten days until weaning, usually at five weeks. A wide range of replacers are on the market from dried milk products with added fat, sugars and minerals to lower-cost vegetable-based powders. All tend to be rather expensive so that the aim is to limit the period of liquid feeding, commensurate with a good growth rate.

Individual penning is ideal, as the intake of each animal can be monitored and controlled. Contaminated buckets are an obvious source of infection, so that thorough cleaning and disinfection are essential at least once a day.

It is often necessary to spend time teaching calves to drink from a bucket, especially those which have been suckling their dam

Having suckled colostrum from its dam, this new-born calf has been placed in a clean, dry individual pen for bucket rearing.

for several days. It may be necessary to get into the pen, stand alongside the calf and back it into a corner. Hold the bucket in one hand, put the other over the calf's head and encourage it to suckle the fingers. Once suckling, the head can be lowered into the bucket, gradually removing the fingers as the calf takes to the milk. Holding the bucket at knee height allows the calf to drink more readily, and hopefully the bucket can soon be placed in the holder on the front of the pen.

Traditionally, milk replacer is fed by bucket twice a day at or around blood temperature. There are, however, some advantages in raising the temperature of the water used for mixing to 40–42°C. The calf then drinks more readily as the flow of saliva is stimulated and the rennet, which is present in the stomach for clotting the milk, is most active at this higher temperature. Consumption should be allowed to build up over the first week until full intake, which for a Friesian calf is 4 litres per day in two equal feeds using 125 g/litre reconstitution. Actual mixing instructions are printed on the bag or an attached leaflet and these should be accurately followed.

Once-a-day milk feeding is an alternative and well-proven

A batch of calves on an all-in/all-out early weaning system. The building will be cleared, disinfected and rested between batches.

system. It involves feeding the same daily level of powder but in a more concentrated form—using 150 g/litre of water. The system is well suited to farms where limited labour is available. Milk feeding on these farms can conveniently be undertaken between milkings, usually in the late morning when other chores have been completed.

Another alternative involves mixing the milk powder with cold water so that the liquid is fed at tap temperature. Calves appear to thrive quite well although they sometimes shiver after a feed. Fuel costs are reduced and it is a system well suited to locations where electricity for water heating is unavailable.

4. Ad-lib Systems

Numerous ad-lib systems of feeding milk replacer have been developed in recent years, particularly to save labour. They involve either various types of dispensing machines or simple storage vessels connected to a series of rubber teats. Rapid growth rates can be achieved from these systems due to the high levels of milk intake coupled with the little-and-often suckling. Animals can be group-housed, thus saving on the cost of individual penning. Some calves need to be trained to use the teat and odd ones that do not take to the system may need to be taken out and bucket fed. Nutritional scours may be seen in the first week; most will clear without treatment but any that have blood-stained faeces should be carefully watched and veterinary advice obtained. Stockmanship is therefore of major importance in ad-lib systems, particularly avoiding the temptation to reduce the frequency of inspection.

Machines can be programmed to provide continuous supplies of milk in the early weeks of rearing, followed by intermittent supplies as the time of weaning approaches. Thorough daily cleaning is essential, especially of the parts which are in contact with the liquid milk.

Specially formulated acidified replacers have been developed which stay fresh for up to three days. They are ideal for use with bulk storage vessels (150–200 litres) as the powder mixes easily with cold water. Some tanks have the teats directly connected so they need to be located in the pen and are therefore difficult to service. Others, more convenient to service, are located in the passageway and supply milk via plastic tubes to a bank of teats

Ad-lib feeding of acidified cold milk. Some time is usually involved teaching young calves to suck the artificial teat.

attached to the pen front.

With all milk feeding systems clean water should be provided from the first week of life. Early-weaning concentrates should also be offered from the second week, these being commonly purchased as small pellets or 'pencils' or a home-mixed ration. The main requirements of this feed are a high nutritional value coupled with palatability. This can be achieved by including such constituents as dried skim milk, molasses or sugar beet pulp. A home mix can be made up as follows:

	%
Flaked maize	50
Crushed Oats	20
Fish meal	10
Linseed cake	10
Dried skim milk	10
Plus minerals and vitamins A and D	

A little soft hay is considered by many rearers to aid rumen development, but others obtain satisfactory performance by allowing calves to take fibre by nibbling straw following daily bedding.

Veal Production

The demand for veal in the United Kingdom is small (0.2 kg per person per year), the majority being utilised by the catering trade. The aim has been to produce 'white veal' by feeding only high-fat milk replacer diets from birth to slaughter at 14–16 weeks of age. It has until recently been common practice to rear veal calves in individual slatted-floored crates without bedding. The crates were located in well-ventilated buildings, and to keep the animals as quiet as possible and reduce fly problems, semi-darkness was maintained. As the animals approached slaughter weight (225 kg) they were too large to turn around in the crates so could only move up and down. These restrictions, together with the complete lack of fibre in the diet, caused considerable public concern so that alternative rearing methods are being developed.

Most rearers now operate with small groups of calves on a straw-bedding system and with ad-lib milk feeding producing veal of most acceptable quality.

Husbandry Tasks

One of the earlier husbandry tasks to be performed is the *removal of supernumerary teats* from heifer calves being reared for replacements. This should be undertaken in the first month of life (by law before three months), cutting off any surplus teats with a pair of sterilised sharp curved surgical scissors. Usually there is little bleeding and it is sufficient to hold a small pad of cotton wood over the site for a minute or so.

Castration of young bull calves destined for beef is necessary unless an intensive feeding system is in operation which can utilise entire males. A common and reliable method of bloodless castration involves the use of rubber rings which by law are restricted in application to the first week of life. The ring is first fitted on to the hooks of the special pliers, and then by squeezing the handles together the ring is stretched enough to fit over the scrotum. Both testicles need to pass through the ring, and on gently releasing the pliers, care has to be taken to avoid nipping the small teats.

An alternative bloodless method is to crush the cord of each testis using a Burdizzo castrator. Each cord in turn has to be pulled to the side of the scrotum, placed in the jaws of the tool and crushed by closing the handles. To ensure success, each cord should be crushed in two different places.

Castration with a knife is very effective but is really a job for the veterinary surgeon. He has by law to undertake the task if the animal is over two months of age and an anaesthetic must be used.

Disbudding is another herdsman's task on farms where calves of horned breeds are reared. The law in this case requires anaesthetic to be used unless the task is done by chemical cauterisation in the first week (not a very reliable method). The most satisfactory method is to cauterise the bud with a special hot iron heated by electricity or by the flame from a bottled gas supply. This task should ideally be undertaken before three weeks of age but there does need to be sufficient growth of horn to hold the cauteriser in position. Local anaesthetic is first given to freeze the nerves running to the horn buds. The site of the injection is in the soft tissue just below the ridge of bone half-way between the eye and the horn bud. The manufacturer's instructions should be followed regarding quantity to use and the injection should be made fairly deep at right angles to the ridge. Hair round the horns should be clipped back to expose the buds during the five to ten minutes required for the anaesthetic to take effect. The effect lasts for up to one hour so that all the group to be disbudded can be injected

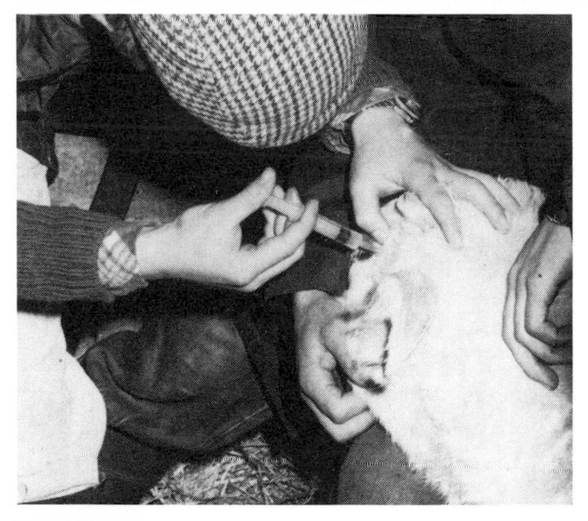

Injecting anaesthetic prior to disbudding. The site of the injection is in the soft tissue below the ridge of bone halfway between the eye and the horn bud.

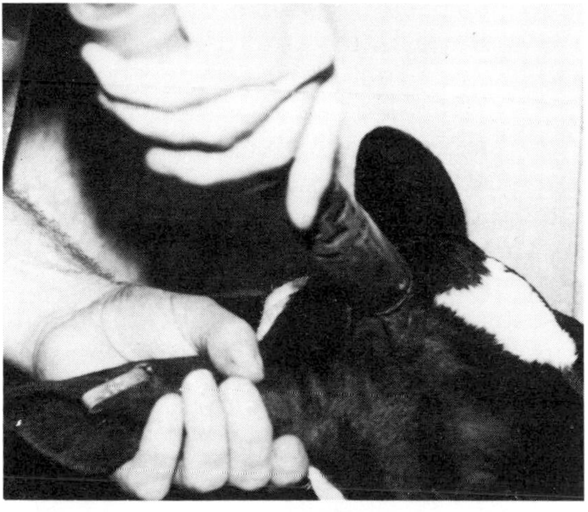

Cauterising the horn bud with a gas-heated iron. It is held on the bud for up to 10 seconds, and the bud is then pushed out with a twisting action.

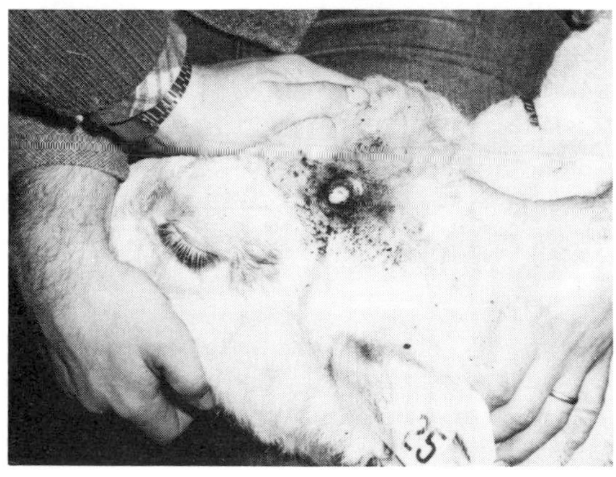

The horn bud cleanly removed.

together. With an assistant firmly holding the head of the calf, the cauteriser is then held on the bud for up to ten seconds, and with a twisting action the bud is pushed out by pressing firmly inwards and downwards with the edge of the cauteriser.

Care should always be taken to avoid fire risks when using gas appliances or electrical extension leads. Keep them dry and out of the reach of calves.

SOME CALF REARING PROBLEMS

Scouring

Although this is the most common disease of the young calf, its effect on the enterprise can be controlled by good husbandry. It can be caused by overfeeding and dirty feeding equipment, but is often associated with infectious bacteria viruses or mycoplasma organisms. The particularly virulent strains, and the ones causing most serious loss, are as follows:

Salmonella This is a bacterial infection which can affect cattle of all ages, but is particularly important as a potential problem in calf rearing. There are a number of strains but *S. typhimurium* (which can also infect man) is the most common. The symptoms are a yellow diarrhoea, high temperature and possibly death within a day. Some calves only suffer from a chronic infection, being unthrifty, whilst others can carry the infection without showing any health problems. These will have high levels of antibodies against the bacteria but at times will shed salmonella in the dung. This may be caused by stress as, for example, in transport, by digestive upset or by poor housing.

Prevention is again preferable to cure, so that if purchasing, buy only strong, healthy-looking calves, restricting the farms of origin to as few as possible—ideally from one source. Such calves need to be treated with care on arrival, gradually introducing milk over the first 24 hours. Vaccination is possible for both *S. typhimurium* and *S. dublin*, but it is advisable not to vaccinate an animal already showing signs of the disease. Salmonella can spread from saliva, so that milk-feeding buckets need to be kept clean.

Virus infections These generally affect calves during the first week of life and cause a watery diarrhoea. The virus destroys the cells lining the intestinal wall, so affecting water reabsorption. The calf is therefore suffering from dehydration so liquid intake is important. Antibiotics have no effect on the virus but they may prevent further bacterial attack.

Scouring caused by rotavirus and coronavirus is reduced when replacement heifers are reared in the same environment as the cows (not on a separate farm) so that antibodies build up and pass to the calf via the colostrum. Bovine viral diarrhoea (BVD) can be a particular problem, especially if the dam of the calf is infected during pregnancy, when the infection passes to the unborn animal.

Respiratory Infections

Calves housed in badly ventilated buildings are liable to suffer from respiratory disease, often called calf pneumonia. The cause is complex; it involves the environment as well as micro-organisms and stress. Improved ventilation is an obvious control method, and vaccines are also being developed.

Abnormal Calves

Occasionally calves are born with hereditary defects such as dropsy or as a bulldog calf. One such abnormality is a freemartin heifer which, as twin to a bull calf, suffers an arrest of sexual development before birth. The AI organisations appreciate notification of abnormalities so that if there is sufficient occurrence future users of the semen can have appropriate warning.

CHAPTER 10

HERD REPLACEMENTS

The average productive life of a dairy cow in the United Kingdom is rather less than five lactations, so that for a 100-cow herd some 20 to 25 heifer replacements are required each year. In suckler beef herds the stress of production is much less than in dairying so that a typical culling rate is 10–15%. In dairy herds a regular annual replacement usually takes place, whereas in a beef herd a group of heifers is more commonly reared or purchased intermittently.

The actual replacement rate in a particular dairy herd is affected by many factors. Market prices of cull cows and replacement heifers often influence replacement decisions, especially in a business where the cash situation is a critical factor. The retention of a number of 'borderline' culls for an additional lactation and either selling the proposed replacement heifers or avoiding purchase has a marked positive effect on cash flow. It must be stressed, however, that such a policy cannot be maintained.

In many herds high levels of enforced culling take place, as disease problems override any decisions by management. Chronic lameness and mastitis are typical problems but the most common one is failure to conceive. This is another major area in which the herdsman is able to make his mark on the profitability of the herd by maintaining a high herd health status and so reducing the level of enforced culling. Unless exceptionally high numbers of replacements are introduced into herds with health management problems, few cows can be culled for low yield, milk quality, conformation faults or behavioural problems. In some circumstances, low-yielders can be justifiably retained if they breed regularly and have no other health problems. Although in some herds three-quartered cows are often retained, any animals which cause persistent problems to herdsmen, such as a continual kicker, should go smartly 'down the road'.

Costings from farm surveys clearly demonstrate that the returns to be obtained from using resources to rear replacements are lower than for milk production, so that high replacement rates create an additional cost burden. The rearing of an above-average requirement of replacements is often claimed to be justifiable when off-land is used which it would not be convenient to use for other purposes. Farmers who do this need to be aware that other resources such as labour, cash and their own management are involved in the enterprise, as well as the land. Now that the risk of diseases such as brucellosis has been eliminated, such inconvenient fields might be more profitably let to other producers or pony owners.

In a herd with a normal culling rate, and allowing for the birth of bull calves, half the herd will be required for breeding heifer calves for replacements. It is an interesting point for discussion as to which cows should be used for this purpose. In a pedigree herd, particularly in one selling surplus animals as breeding stock, the highest yielders and cows with superior conformation will normally be used regardless of date of calving. In strictly commercial milk-producing herds, especially in those with a tight calving pattern, it will be the first half of the herd to calve which should breed the replacements. National Milk Records now rank animals in the herd in terms of their relative breeding merit, but this does not take calving date into account.

With price incentives now operating for summer milk production, there is added emphasis on rearing early autumn-born heifer calves to join the herd at 22–24 months of age. This is a considerable challenge to stockmen, and as can be seen below, it is possible with care and adequate resources.

Another method of obtaining replacement calves, and at the same time increasing genetic improvement in the herd, is to breed heifers from heifers, i.e. serving maiden heifers to the pure dairy bull. Selection of an appropriate sire is vital, particularly in respect of low reported incidence of difficult calvings.

Rearing Policies

There are three main alternative methods of obtaining replacement stock, namely to purchase, to home rear and to employ a contract rearer.

The purchase of down-calving heifers is the usual method for a 'flying' herd but is also used periodically by a wide range of dairy and beef farmers. Animals can be obtained of a required size, age, calving date and particularly of required conformation. Although considerable lump sums of cash are required (with some risk as to the actual amount), money is not tied up in a rearing enterprise.

Neither land nor management time is involved and in the present situation disease risk is minimal, especially if none of the heifers has calved before arrival. Buying 'on the farm' rather than in markets reduces the disease risk, with leptospirosis and BSE now being added to the usual risks such as mastitis. Travelling to dispersal sales can be expensive in time and money, and the cost of transporting the animals home is also a factor to be considered. Some farmers and their herdsmen get considerable satisfaction from studying performance data of sires and dams as well as from attending the sales to select their replacements.

Many more prefer to rear their own heifers, to avoid disease risk and to have the pleasure of seeing the result of planned matings develop into full members of the herd. As mentioned above, considerable resources are involved and few producers are aware of the actual full cost of producing a home-bred down-calver.

Contract rearing, although not widely practised in the industry, does offer considerable advantages to the dairy farmer who is short of land and perhaps labour. He is able to have the calves reared from his own cows at a known cost as well as with minimal involvement of his time. Disease risks depend upon the number of other producers the rearer is dealing with and the quality of the down-calver, to a large extent reflecting the skill of the rearer. The system offers the rearer a regular income with reduced capital outlay. It is well suited to a grassland farm which has perhaps discontinued milk production or to an arable farm as a means of utilising rotational grassland without capital injection. A carefully worded contract needs to be drawn up, to include such detail as which party covers the cost of veterinary bills and anthelmintics.

In beef herds the source of replacements is somewhat different as many suckler cows are commonly cross-bred animals, so are rarely born in the home herd. Beef × dairy heifers are very popular especially the Hereford × Friesian which can be readily purchased at almost any stage from the young calf to the down-calver. Other more specialised crosses such as the Blue-Grey can be purchased or produced at home by the herd having a few pure Galloway cows which are served or inseminated with a White Shorthorn bull.

Age at Calving

The next factor to be considered by farmers rearing heifers or having them contract reared is the optimum age of calving. It is a most important factor because not only does it determine the resources involved in rearing, but it also affects the calving pattern and long-term production of the herd. There is considerable evidence to indicate that so long as satisfactory growth takes place, heifers calving at two years of age rather than 2½ or three years have a higher lifetime production. Although the yield is rather less in the first lactation, younger-calved heifers are, on average, retained in the herd for a longer period of time, so producing more milk. Many farmers and their stockmen have reservations about early calving but tend to confuse the interaction between age and weight, knowing that small heifers usually have unsatisfactory performance.

Research findings show that it is unsatisfactory to grow heifers on a very high plane of nutrition to calve at less than 22 months, dystokia being a major problem. With good feeding and management it is a realistic target to produce a heifer weighing 500 kg just before calving at two years of age.

An associated factor to be considered along with the age at calving is the calving pattern of the herd that the heifers are to join. If there is a tight seasonal pattern (most animals calving within two months) as in a few dairy and beef herds, then to maintain the pattern, heifers need to calve at either two or three years old. In herds with a spread of calving dates there is a wider range of options for age at calving, especially if no attempt is being made to tighten the pattern. The calf- and heifer-rearing programmes are complicated on farms which retain heifer calves over an extended period. Unless numerous groups can be formed, competition is a problem in a group with widely varying sizes of heifers.

Owners of some high-yielding herds prefer to calve their heifers as a group, several weeks before the main herd, so as to allow staff more time to settle the heifers into lactation satisfactorily and so lift peak yield as well as total lactation production. Some delay is then allowed in first service which, together with a rather longer dry period, allows the heifers to continue growing prior to their second calvings.

Whichever rearing system is adopted, it is advisable to have a definite plan for heifers to calve in a given month. Once this date has been decided, growth rates to achieve that target can be assessed. The major task is then to ensure that nutrition, health and environment are maintained at a satisfactory level for adequate growth. Target growth rates need to be achieved at every stage of rearing, as the need for accelerated growth rates in subsequent periods is not only expensive (if only because of the need for concentrates), but it may also lead to over-fat heifers and to more dystokia.

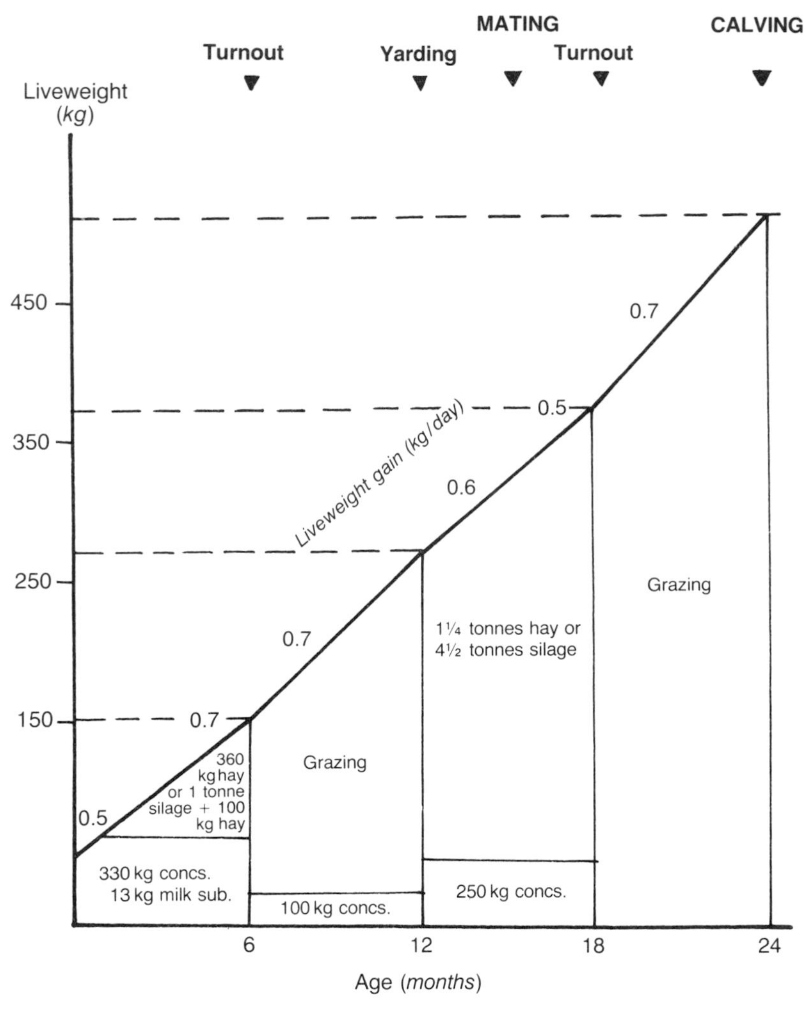

Total requirements: 13 kg milk substitute
680 kg concentrates
5.5 tonnes silage + 100 kg hay, or 1.9 tonnes hay

Note: Feed quantities are approximate.

Source: MMB/MLC.

Figure 11. Targets for autumn-born calves to calve at two years.

Yet again there is the need for first-class stockmanship to observe and monitor performance so that satisfactory growth is obtained throughout rearing. The role of management is also important in providing the necessary resources, particularly quality forages and adequate housing and handling facilities. Figure 11 indicates the target liveweight gains for the heifer calving in the autumn at two years old. This is the most common type of heifer being reared and so will be used as an example in describing the details of a rearing system in the remainder of this chapter.

Calf Rearing Stage

In the case of the autumn-born calf, targets are more easily achieved with the earlier-born arrival, that is if they are born in September rather than December. Older calves at pasture during their first grazing season have a higher appetite and so can make better use of grass.

During the calf stage, an early-weaning system is ideal, using bucket rearing or an ad-lib system of milk feeding as described in the last chapter.

First Grazing Season

Few difficulties should arise until turn-out, which is ideally at six months of age and 150 kg weight, when the major challenge to herdsmen begins. Where husk is a problem, vaccination will need to take place some six and two weeks before turn-out. A good leafy sward will maintain growth rate but over-lush or stemmy grass is less satisfactory. Concentrate supplementation in the early weeks at grass (up to 1.5 kg per day) will be beneficial and may need to be reintroduced in very dry periods but particularly at the end of the season as forage quality declines. Newly sown leys provide an ideal feed, being free from parasitic worms, but older leys or permanent grass have often to be used. Grazing management is therefore most important in such situations, allowing the young heifers to satisfy their appetites with nutritious grass. A move to 'clean' aftermath in late July is of benefit, possibly after second-cut silage, or even to a pasture grazed earlier by another class of animal, such as ewes and lambs. Dosing with anthelmintics in July has been to date the recommendation for worm control, but if improved long-acting, sustained-release boluses are 'put' in the rumen before turn-out,

Young calves at grass.

moving the animals back to a handling area in mid season is avoided.

Set-stocking and paddock grazing are widely used systems for calf grazing; good results have also been obtained with the leader-follower system. This involves a rotational system of grazing with the young calves being one paddock ahead of a group of older heifers. The calves selectively graze, with the older animals following to utilise the remaining herbage. Stocking rates are dependent upon the level of nitrogen fertiliser applied so that if a high level of 300 kg per hectare is used, two livestock units (two young heifers plus two in-calf animals) can be carried per hectare.

Regular weighing of the animals when at pasture is recommended so long as the handling facilities are close by, minimising the disturbance which is created. A move to better grazing or even concentrate feeding may be required if target growth is not being achieved. As with the milking herd, buffer feeding is a practical alternative to maintain performance at times when pasture growth is unsatisfactory.

Warts on the teats are sometimes a problem to yearling and in-calf heifers, caused by a virus thought to be transmitted by flies—hence the need for a fly control programme, e.g. use of fly-repellent ear-tags. A serious outbreak of warts may justify the vet taking sample material which can be used to produce a specific vaccine, although a high level of control is seldom achieved.

Yearlings—Housing and Service

The changeover period to housing at the end of the first grazing season is another critical one for avoiding any check to growth. Introducing concentrate as well as a little hay or silage at pasture will help the rumen flora to adjust to the change of diet.

Target weight at service for a Friesian heifer is 330 kg and for early autumn calving the service has to take place in December, which does not allow much time for recovery from any loss of weight on housing. High-quality conserved forages are essential at this time, together with concentrates at least until service.

The choice of bull to use on heifers is influenced by the need to avoid difficult calvings but at the same time to produce a valuable beef cross calf or even a replacement heifer—if the herd is being expanded. Large beef breeds should be avoided. The Aberdeen Angus will produce small calves with minimal dystokia but lower growth potential than the Hereford, which is widely used for this purpose. Natural service prevents the need for oestrus detection which can be a difficult task in the winter months, especially in isolated locations without electricity supply.

The use of prostaglandins has proved to be a convenient method of bringing a group of heifers into oestrus together so that AI can be used without oestrus detection. Satisfactory conception rates appear to be achieved when the animals are in a rising plane of nutrition at service time. The usual system is to arrange for the vet to inject prostaglandin intramuscularly on two occasions with an 11-day interval. Insemination is then carried out at both 72 and 90 hours (days three and four) after the second injection. Special arrangements need to be made with the AI centre to book the necessary semen but more especially the inseminator time. Some saving in cost is achieved if only one insemination at 80 hours is used, but then a 'clean up' bull needs to be used for repeat breeding.

Apart from a routine pregnancy check some two months after the end of mating, minimal management input should be required until turn-out. Target weight gain after conception is

Yearling heifers at the drinking trough. Note the glossy coats indicating good health and growth.

Yarded heifers.

only 0.5 kg per day, which should be achieved with a high-quality forage diet.

Second Grazing Season

During the second grazing season weight gains of 0.7 kg per day should also be obtained without difficulty if good grassland management is practised. As the weeks to calving approach, a watch on body condition is required aiming at a level of 3.0 to 3.5. Overfeeding at this time has to be avoided; otherwise large calves will be the result and there will be considerable dystokia problems.

Frequent inspection is required at this time, especially to watch for any signs of summer mastitis. Offering a small quantity of concentrates at each inspection helps the stockman to be able to handle the heifers and is especially useful when checking for the onset of calving. If the down-calvers are located near the milking facilities it is advantageous to allow the group to walk periodically through the parlour to get used to the new environment. After one or two visits the pump and pulsators can be switched on so that when a heifer calves and has to be milked she is less likely to cause problems in the parlour.

Replacement heifers are the future members of the herd so that it is well worthwhile giving care and attention not only during the rearing period but especially at calving time.

CHAPTER 11

BREEDING, AI AND BULLS

A major objective in profitable dairying is to get 85–90% of the herd in calf each season with a calving interval of no more than 370 days. Survey results show that failure to conceive is the most important reason for culling dairy cows of any age, apart from very old animals. Suckler beef cows also need to calve regularly in order to maintain a tight calving pattern and so ease feeding and management problems. Good herd feeding, housing, freedom from disease and bull selection are factors which contribute to successful breeding but there is no doubt that one of the key factors is stockmanship. It is generally agreed that only a small percentage of cow fertility is influenced by heredity, the remainder being in the hands of the management and herdsmen.

Regular calving increases the output of milk and calves; it also reduces veterinary costs and the loss of good cows through slaughter, with delays in conception having a marked economic effect on the enterprise. As the levels of feeding and management in a herd improve it becomes increasingly important to have animals with a high genetic potential, so that the selection of sires is another subject in which herdsmen should be closely involved. Breeding is a long-term process with no guaranteed results, although one hopes that by careful selection of bulls, heifers will turn out to be better cows than their dams. However, this will depend on how well the genes of the sire and dam 'nick' together. There is little doubt that in the future, computer programmes will increasingly be used as a method of sire and dam selection.

To assist in selecting the best cows in the herd in terms of their potential to transmit the ability to yield fat and protein, cow genetic indices (CGIs) are now calculated and published. Currently these are available only for pedigree animals but it is hoped that there will soon be wider coverage. The CGI for a cow is a number between 500 and 1,000 (the higher the better) based upon four sources of information:

- The cow's own production
- The ICC proof of her sire (see below)
- The CGI of her dam
- The average genetic level of the herd

Although as stated above, such factors as calving date, conformation and temperament need to be taken into account in deciding which cows should be given the opportunity to breed replacements, it is a great help to have a genetic ranking for the herd with CGIs. The system is also ideal for the AI organisations to identify the 'top' cows nationally as breeders of future AI sires.

Some essential data for breeding decisions will be automatically gathered by the milk recording process but herdsmen will continue to have an important role in providing information on such factors as temperament and ease of milking.

This chapter deals with the role of the herdsman in managing fertility. It first deals with the female aspects of reproduction, then with AI, DIY AI, ova transplants and concludes with bull-rearing and management.

Infertility

Some infertility problems arise from infections of the reproductive tract of the cow. These are usually localised non-specific types of infection, although others such as venereal disease are spread by natural service. Brucellosis was, until its eradication, a major infection of the genital tract which seriously affected fertility. Animals which have difficult calvings, torn or bruised tissue or a retained placenta commonly develop infection of one or more sections of the breeding tract. These animals produce abnormal discharges which are foul smelling, and because the pus is a browny or yellowy or white colour, it is often known as 'whites'. This problem requires quick spotting and prompt treatment by the vet using uterine antibiotics.

Correct diagnosis is another important skill required of the herdsman. 'Whites' should not be confused with the slightly cloudy, often blood-tinged discharge which sometimes appears two or three days after oestrus. This is quite normal and even if the cow has been served, it does not mean that she has failed to conceive.

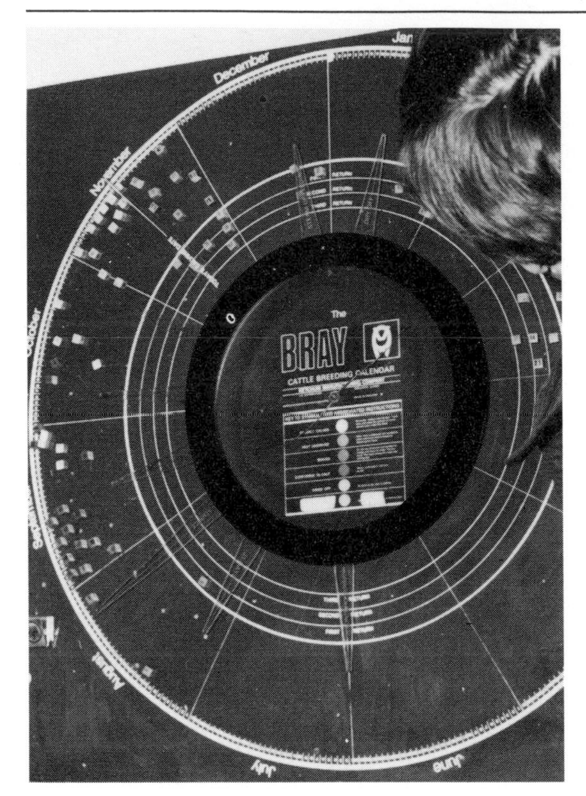

A rotary breeding board used to predict forthcoming events such as calving, oestrus or drying off.

The cow standing to be mounted is in oestrus and ready to be served or inseminated.

Post-calving Management

A cow which calves and cleanses normally and is well fed in early lactation should come into oestrus within three or four weeks after calving (the date needs to be recorded). The uterus is stimulated back into normal condition by oestrus, due to an increased flow of blood in this region of the body. Cows not seen in oestrus by 40 days should be examined by the vet, this being particularly important at the start of the breeding season. The length of gestation is on average 280 days so that to produce a calf at exactly a one-year interval, the animal would need to conceive on the 85th day of lactation. A useful guide in practice is to serve the cow at the next oestrus after 50 days in lactation, assuming she has already had a previous oestrus, because attempting to serve at the first oestrus reduces the chance of conception. This is thought to be caused by the fact that the uterus has not returned to normal after the previous pregnancy and is therefore not ready to accept the embryo. Following a difficult calving or retained placenta it is preferable to delay service to allow time for full treatment of any infection.

Research has shown that the chance of conception is improved if the cow at service has started to regain the weight she will usually have lost in the early weeks of lactation. Obtaining accurate body weight is impractical on most farms so that condition scoring as described earlier is a practical alternative, the 'guide' score at service being 2.0.

Other research has shown improved conception from cows well fed, especially in energy terms, at the time of service. The diet needs to provide adequate energy for maintenance, milk production and 10–20 extra MJ for liveweight gain.

Breeding Records

Chapter 15 deals with the collection, analysis and use of records, including those required for herd breeding. It is, however, worth stressing at this point that an efficient recording system is a key factor in successful breeding. In beef herds using natural service, the only data required are the dates when the bulls were turned in and taken out, whereas with a dairy herd a comprehensive system of recording is justified.

Heat Detection and Service

In herds which rely solely on natural service there is less need for herdsmen to be fully skilled in heat detection. However, even in herds where cows have to be taken to the bull, time can be saved by only presenting the ones that are well in heat and will stand to be served. In the majority of herds, i.e. those using AI, it is a vital part of successful breeding management that herdsmen, and in fact all farm staff, are skilled and effective in heat detection. It is not a simple task, as studies have shown a very wide range in the length of time that individual animals are in oestrus—mature cows from only half an hour to 25 hours, with maiden heifers seldom for more than 12 hours. A few animals will understandably be missed if the oestrus occurs during the night hours, but the aim should be to identify 85% of predicted heats (a simple calculation which can be undertaken weekly or even monthly). During the breeding season, three or even four periods of at least half an hour need to be set aside solely for this purpose. These sessions should preferably be at times when the herd is settled and is not expecting to be fed, moved or milked. One of the key times is late at night, so that adequate lighting should be available in winter housing.

A cow is right for immediate insemination only when she stands to be mounted. In herds practising batch calving, or in large herds where cows are yarded by calving dates, several animals may be in oestrus at the same time and form a group, making accurate identification most challenging. It is also necessary for stockmen to recognise the behaviour of animals coming into and going off heat so as to arrange appropriate timing of insemination. In the majority of situations, the inseminator will come to the farm following a phone call, at approximately the same time each day. It is necessary, therefore, to have an appropriate cut-off time, after which any cow seen coming into heat is saved for the following day. When 'Do it Yourself AI' (DIY AI) is in operation, there is the opportunity for insemination to be carried out at the optimum time, i.e. when the cow is standing for mounting, although in practice the job is usually fitted in as and when convenient.

Numerous heat detection aids are available to help identify oestrus in problem cows. These include the Kamar heat mount detector which is stuck with glue on the sacrum of the cow, just in front of the tail head, changing colour when she is mounted by another cow. A more recent development is a paint-like paste which is put on at the same place just in front of the tail head and which is rubbed off when the cow is mounted. It appears to last for some two weeks before it requires topping-up.

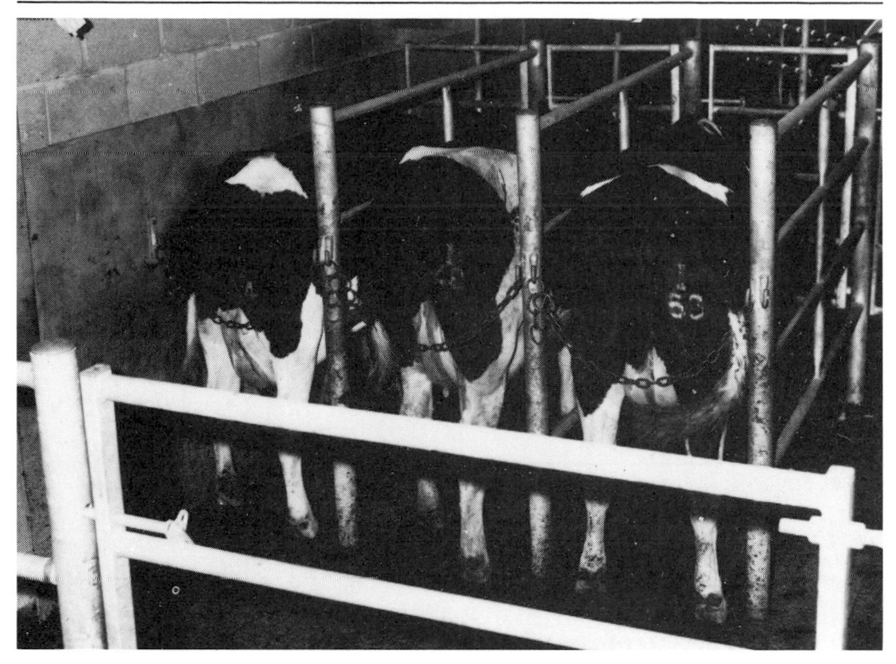

Purpose built AI stalls adjacent to the exit from the milking parlour.

Artificial insemination.

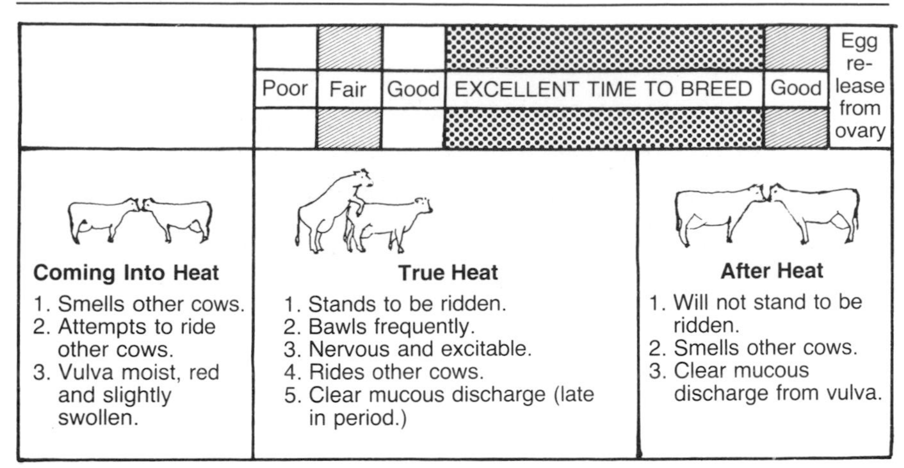

Source: MMB.

Figure 12. The best time for breeding.

Progesterone testing of milk is increasingly popular as an aid to fertility management. It is used as a method of pregnancy testing (see below) and also of oestrus detection. The basis of the test is to measure the varying level of progesterone—which is almost zero at oestrus, increasing to a peak 17–20 days later and falling suddenly again at the next oestrus. (The fall does not occur if the animal is pregnant.)

A small representative sample of milk is taken at milking time from the collecting jar into a clean container marked with the cow's number. If the sample is taken direct from the udder, avoid foremilk and ensure the sample is made up from all four quarters. Samples not tested at once can be stored for 24 hours in a refrigerator (which is also required to store the test kit). The system works on colour change of the sample comparing with a 'standard'; full instructions are provided with each kit and to obtain a high level of accuracy these need to be followed with precision. Many herdsmen consider this method too fiddly, whilst others 'speak nicely' to the Farm Secretary and so obtain skilled assistance!

In some circumstances, as for example with heifers wintering in an 'away' location, it may be preferable to use prostaglandins to synchronise heat within the group. This enables the insemination to be undertaken at pre-arranged times without the need to identify oestrus (two inseminations being usually carried out on successive days). It should be remembered that these products are not 'fertility drugs' and good conception results will only be achieved if the cattle are cycling normally, receiving an adequate diet and in a suitable condition for breeding. Another commonly used method of bringing a cow into oestrus is the progesterone releasing intravaginal device (PRID). This involves using a special applicator to insert a flexible coil impregnated with the hormone progesterone into the cow's uterus. The PRID is removed after 12 days and the cow should come into oestrus two or three days later. 'Sweeper' bulls are normally used to serve those animals which do not hold to the inseminations.

Another valuable aid to good conception is to ensure that the animals are securely and safely restrained when presented for AI. It is usually convenient for the inseminator to 'operate' on cows tied in a cowshed, retained in a crush or yoke or even in an abreast parlour, but not so convenient in a herringbone. Many dairy layouts incorporate a set of purpose-built stalls into the exit from the parlour which are ideal for AI as well as pregnancy diagnosis and many other veterinary tasks.

If no one will be around when the inseminator calls, it is essential to identify each animal clearly by tying on a label (supplies available from the local centre) or chalking the details on a board above the cow. It is also necessary to leave soap, water and a clean towel for the inseminator to wash down and carry out his routine disinfection. A bin should also be available for disposal of used straws and soiled plastic gloves. It pays to cooperate fully with the inseminators at all times, so that they look forward to coming to the dairy and helping achieve good conception results.

DIY AI

A number of owners, particularly those with larger herds, have purchased semen storage flasks for the farm and have arranged for members of their staff to undergo training in insemination. (At least two people need to be trained to cover for sickness and holiday periods.) Some saving in cost is claimed and it also creates additional interests for well motivated herdsmen. It is, of course, advantageous in severe weather conditions or in a foot-and-mouth outbreak when the inseminators cannot get to the farm. As mentioned above, it is possible to inseminate at the optimum time and even to double-serve in certain cases. Conception rates are variable and depend considerably on the skills of the individuals doing the job, with retraining sessions being a necessity before each breeding season. There is considerable extra paperwork involved and regulations to be followed, including the provision of a locked container for the storage flask. In a herd where block-calving is undertaken, so leading to several months when no inseminations are made, it is beneficial for staff to undergo a short refresher course to sharpen up their technique before the new breeding season.

Ova and Embryo Transplants

This is a relatively new technique which due to cost has been restricted to very high-value animals of both dairy and beef breeds. The donor cow has to be transferred to a specialised centre for the operator to recover fertilised eggs and implantation into the recipient cows or heifers also takes place there.

Developments are however underway to enable the implantation of fertilised ova to replace AI in commercial breeding. It will also be possible to specify the sex of calf, which will benefit the beef sector just as much as dairying. The production of replacement heifers will come then from 20–25% of the herd, the remainder being available for implantation with three-quarter-bred, or even pure-bred ova to produce high-potential male beef calves.

Pregnancy Diagnosis (PD)

It is advantageous to arrange for the vet to carry out pregnancy diagnosis (PD) as soon as possible after service. For many this will be after some 90 days, although a number of highly experienced practitioners can make accurate assessment at 40–50 days.

The milk progesterone test, as outlined above, can be used between the eighteenth and twenty-fourth day following service. Accuracy is over 95% for negative results and over 85% for positives.

Selecting Sires

This is a challenging task. Nevertheless, undertaking or being involved with the management in the selection of sires to use on both cows and maiden heifers can be a most rewarding aspect of a herdsman's job. Advice on this subject is available from a wide range of sources but those who have to take decisions are strongly recommended to contact their AI Centre Manager or ADAS Livestock Advisers for the latest information. A wide range of bulls

GRAZING SYSTEMS

Strip grazing a recently established ley.▲

◀ Set-stocking with permanent pasture.

ɪrsey cows receiving straw as a buffer ed in a dry summer.▶

ɔving irrigation equipment. This task n often be conveniently fitted into the rdsman's day during the less busy mmer months.▼

SILAGE MAKING

Commencing to flat-roll. An ideal ▶ opportunity to pick up any 'foreign bodies' that could damage the forage harvester.

A contractor operating a self-propelled forage harvester to pick-up, chop and side-load wilted grass.▼

▲Filling a concrete walled silo with high-quality material. Note the black plastic side sheeting which will form a vital part of the sealing process on completion of filling.

The same clamp now▶ filled and effectively sealed. The close-fitting tyres hold the top sheeting firmly in place.

DAIRY COW HOUSING

A well-designed and managed cow kennel unit. ▶

A new design of cubicle division to provide optimum space for the animals.▼

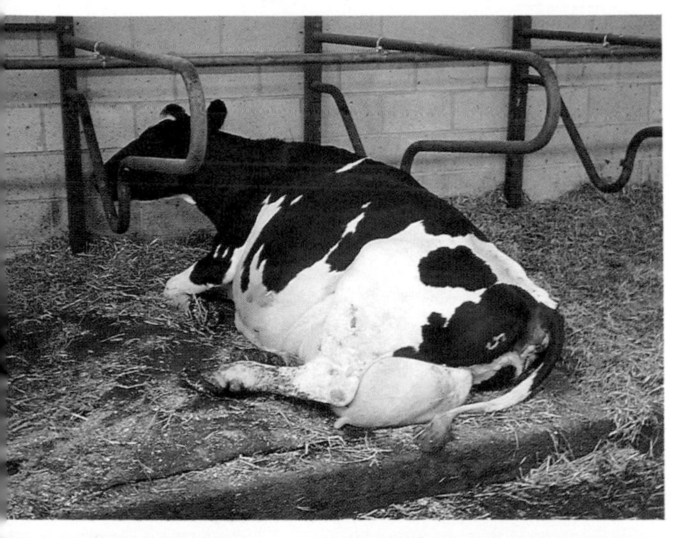

◀ Bedding-up a cubicle house using good clean straw from a large bale.

Moving slurry into a storage area with a tractor-mounted scraper.▼

HERD HEALTH

The result of good health management during the rearing of a replacement heifer to calve at 24 months weighing 500 kg.▲

Mastitis control. Infusing a tube of antibiotic into an infected quarter following thorough stripping and disinfection of the teat end.▶

Lifting a downer cow with a specially designed harness. Note that the back legs of the cow are restrained by shackles to prevent her 'doing the splits'.▼

FOOT CARE

Loading a cow into a rotating crush. Note the two straps around the cow which take her weight before the crush is turned through 90° (using tractor ◀ hydraulics).

A professional foot trimmer paring overgrown hooves using sharp, well-maintained clippers.▶

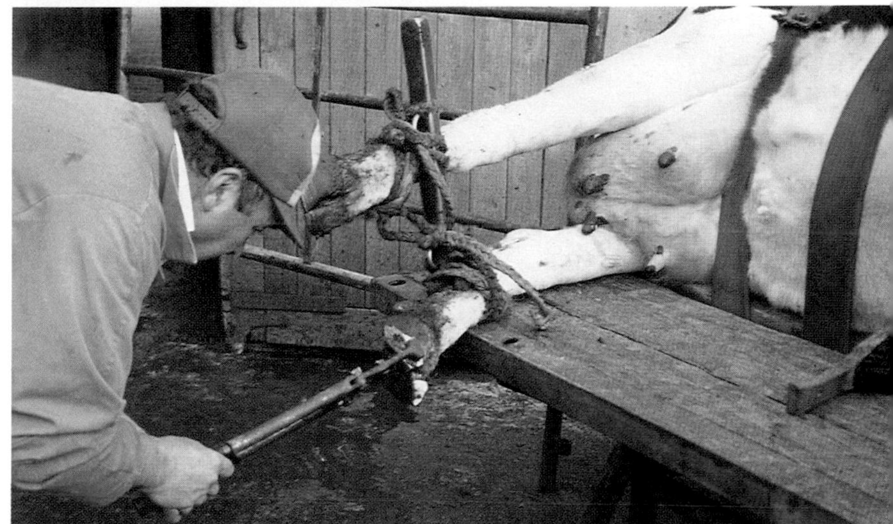

A final trim with a power-operated abrasive disc.▼

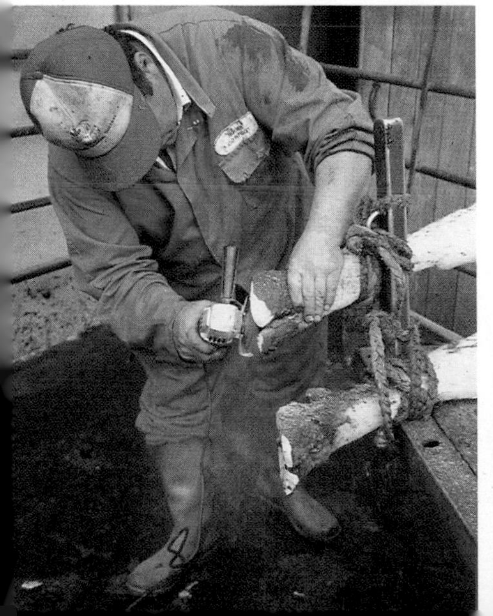

Gluing a temporary wooden block to the good claw in order to protect the tender infected one from contact with the ground and so assist healing.▼

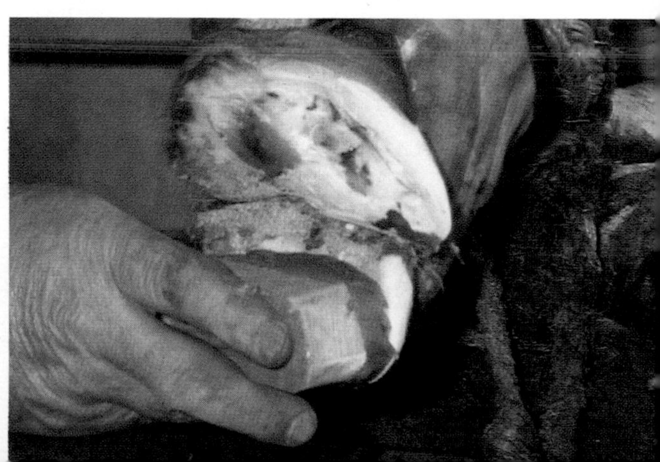

OFFICE JOBS

Inputting the day's records into the farm-based computer. ▶

The farm secretary using the milk progesterone test as a valuable aid to fertility management.▲

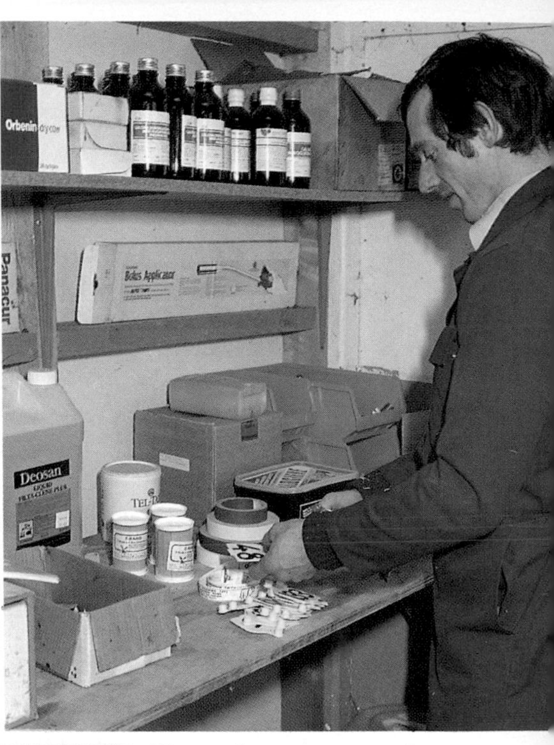

Checking stocks in the dairy stores. Another area of the herdsman's job which involves record keeping.▶

Discussing straw requirements, transport and storage arrangements for the new season with ◀ colleagues.

CALF REARING

Transferring the new-born calf from the field using a tractor linkbox. If all goes to plan, the dam ◀ follows.

Spraying the navel with antibiotic to prevent infection entering the body by this route.▼

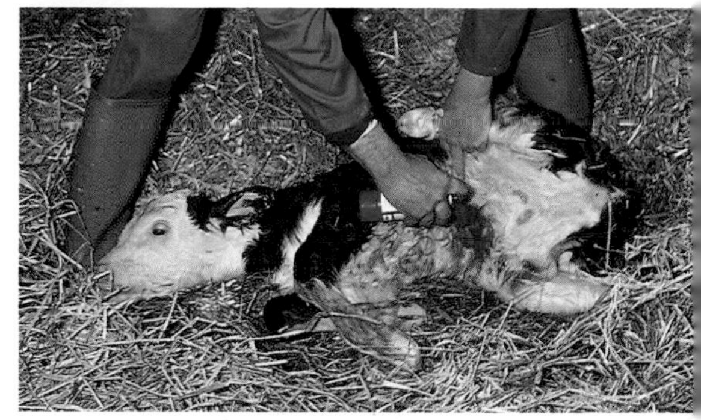

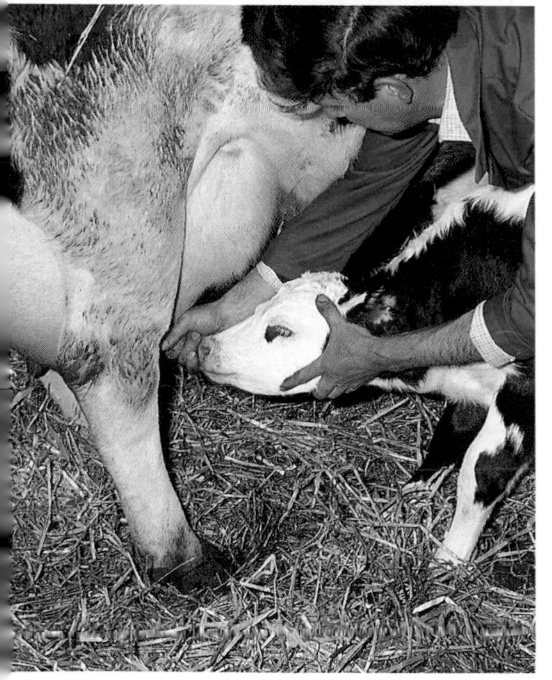

Encouraging the newly born calf to suckle ts dam and obtain vital colostrum. A task equiring considerable patience.

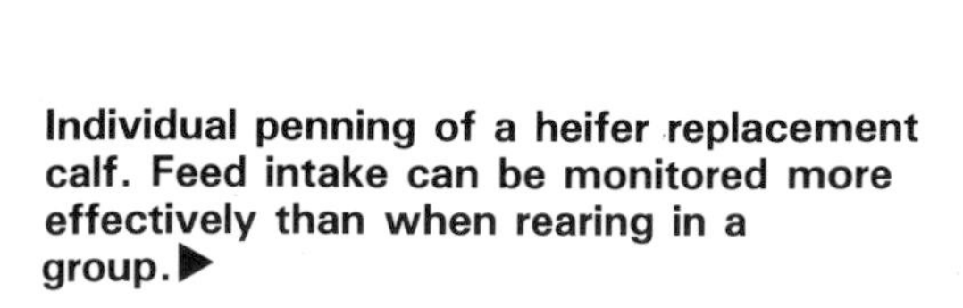

Individual penning of a heifer replacement calf. Feed intake can be monitored more effectively than when rearing in a group.▶

FIELD WORK

Fertiliser application involving effective calibration, accurate driving and a thorough washing of the machine at the completion of ◀the job.

Pasture topping. This improves the quality of the herbage before the next grazing and also presents a good opportunity for a break from the regular routin ◀ in the buildings.

Manure spreading. ▲

Replacing a broken post in an existing fence line. It is preferable to have an assistant available even when using this hydraulic post-driver.▶

of many breeds are readily available from Genus, independent cattle breeding centres and breeding companies. The bulls are of varying potential, proven ability and, of course, price!

In a flying herd, where all replacements are purchased, a beef sire which produces high-value calves, but without undue risk of calving problems, will produce better results. In a stable-sized dairy herd rearing replacements, it may be preferable to breed the best 60% of the herd to top proven sires and the remainder to beef bulls. In selecting proven sires, the production figures of his progeny are of major consideration but so are confirmation, reliability of the data, his fertility and the value for money of the semen.

In order to evaluate genetic merit, the AI organisations have developed sophisticated techniques using improved contemporary comparisons (ICCs). To eliminate the influence of feeding and management which account for much of the difference in performance between herds, daughters of a bull are compared with daughters of other bulls calving in the same herd at the same time. Lactation records are corrected to allow for the age and month of calving of the heifer. Only those calving in the same four-month period are classed as contemporaries. An ICC is the average amount that a bull would be expected to transmit to his progeny related to a bull ranked zero. In selecting a sire, as long as he has higher genetic merit than the cow he is to be bred to, improvement in performance is to be hoped for. It is preferable to select for weight of fat and protein and not for percentages. The highest income is obtained from animals which produce the greatest weight of butter fat and protein. If the herd figures for fat are near the penalty zone, then it would be wise to select a sire with high percentage fat figures. Included in each set of data for a bull is the 'weighting', which is a figure indicating the reliability of the ICC. As more daughters complete lactations, the weighting increases and only bulls with a weighting over 30 can be used with confidence. It is preferable for the daughters to have been milked in at least ten herds with no more than 25% in any two herds.

Conformation ratings are provided by each of the numerous breeding organisations, often on a differing basis, so that valid comparisons can only be made between bulls evaluated by the same scheme, i.e. Genus or British Holstein Society. The main conformation factors are udder shape and attachment, teat shape and placement as well as legs and feet. Most organisations provide a useful brief summary of the typical daughters of a bull, e.g.

A cattle progeny-test panel involved in a 'workshop' session to evaluate a new method of classifying conformation.

large cattle, Canadian-type, with good udders, legs and feet just above average.

Note that these conformation figures are only averages so there is no guarantee of the results! It is hoped that in the near future a universally agreed system of classification will be in operation, particularly for breeders wishing to select and use 'genetics' from other countries.

Rearing Bulls

It has been common practice to single-suckle bull calves being reared for service and in some beef breeds even to foster the calf on to a dairy cow. Bucket rearing, however, provides an entirely satisfactory start for a bull calf, which may, with advantage, be weaned a few weeks later than a steer or heifer (at seven–eight weeks). A bull should be handled frequently from a young age and taught to lead on a halter and be spoken to, so that his confidence is gained. A ring needs to be fitted by nine months of age, usually using a self-piercing ring into the septum of the nose, but it is preferable to bore a hole with a punch so as to locate the ring more accurately. When handling an older bull it is essential to use a staff and to keep his head held up high. After sexual maturity, at nine–ten months of age, it is necessary to pen bulls separately but also to maintain regular exercise and handling on hard ground to help keep the feet in good shape.

Housing for Bulls

With older bulls, especially of the dairy breeds, it is essential to have properly designed boxes with adjacent exercise yards meeting HSE standards. The internal walls need to be smooth to avoid damage to the bull and a feeding hatch incorporated to make feeding easy and safe. Attached to the exercise yard should be a service crate where the cow to be served can be let in through a gate which shuts off the crate from the bull yard. In order for the stockman to clean out the pen in safety, either a heavy sliding door can be used to separate the box from the yard, or a feeding stall can be installed with a guillotine gate at the back to hold in the bull. It is necessary to be able to catch the bull; this can be done at the manger if a yoke is fitted (not really practical with a horned animal) or by having a short length of fine rope fitted to the ring to catch him while he is feeding.

Feeding the Bull

The overall aim is to keep the bull fit without being fat. Many young beef bulls being prepared for sale are allowed to get too heavy and may have to be slimmed down before working effectively. Unless the bull is working hard a forage-based diet may be adequate, hay usually being most convenient; 1.3 kg per 100 kg liveweight is a guide, plus 2 to 3 kg concentrates per day when in active use. During the summer it may be possible to tether the bull on pasture. This system affords exercise and avoids boredom, but he will need some shelter from the hot sun.

Working the Bull

A young bull should not be overworked in the early part of his life. He can be turned into a yard with a small group of heifers to play around and arouse his sexual desires. Once he has successfully served a few heifers, he can then be used on bigger cows. If a very big animal is to be served it may be necessary to

Handling a difficult bull on two poles.

A bull pen located alongside the cow yard. This is a valuable aid to oestrus detection as well as being convenient when natural service is undertaken.

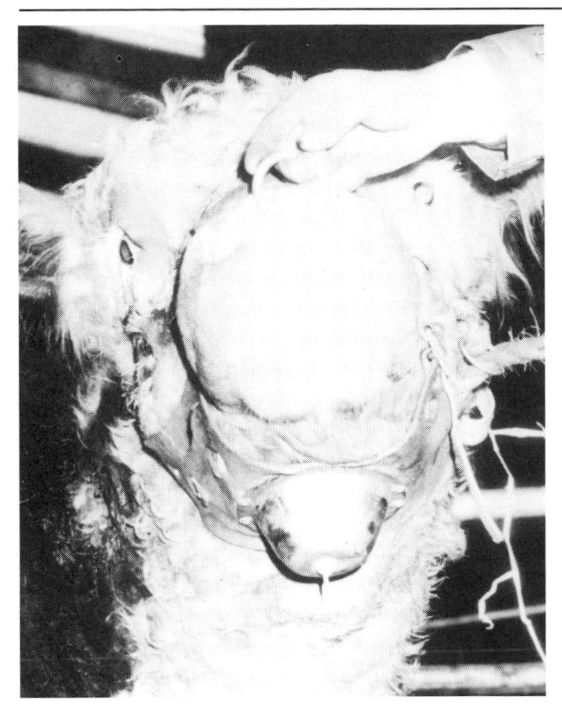

A vasectomised bull fitted with a chinball marker. This is another useful aid to oestrus detection.

dig a pit for her hind-legs to stand in. At all times slippery floors should be avoided. Some older bulls become slow in service owing to lameness in a hind foot or fear of pain or injury, or even strange surroundings. It may help a bull to be turned into a safe yard with a group of cows to build up his confidence. During the 'off-season', loneliness and confinement often lead to a deterioration of temper, and when the new breeding season commences he may need to serve a cow more than once as the semen will initially be of low fertility. When a beef bull is run with the herd he can easily be overworked, so it may be necessary to have at least two bulls so that they can be switched around and each allowed a rest to freshen up. In high-yielding herds where concentrates are fed with other feeds out of the parlour, there is a danger that bulls which run with the herd will get too fat. Not only does their work suffer, but overweight bulls can damage parlour gates and stalls if allowed in with the cows. A bull which is producing much semen and showing excessive signs of libido is not necessarily fertile. If it is found that cows are not holding, then the vet should be asked to take a semen sample for quality testing. It is advantageous before buying a bull to have his semen tested. Following satisfactory results, it is wise to use him on a few heifers and have pregnancy diagnosis results before using him more widely in the herd.

Another reason for having a bull on the farm is to help with oestrus detection of AI. Positioning the bull in a pen alongside the cow yard or adjacent to the exit route from the parlour attracts the cows in oestrus. Some farms use vasectomised bulls (teasers) to aid oestrus detection, usually fitting them with a chinball marking device on a head-collar. When mounting a cow, the bull leaves a paint mark on her back; colours are usually changed every three weeks. A problem can arise with a teaser in a large group of cows (as it can with an entire bull) when several animals are on heat together and one misses being marked.

Bull Health

One of the main difficulties with natural service is the risk of transmitting infection within the herd. Genital vibriosis is spread in this way and can be a cause of abortion at four to five months. There is a particular risk with hiring or borrowing a bull so that treatment for possible vibriosis should first be arranged. It is important to keep the feet of a bull in good shape and it may be necessary to cast him or have cattle stocks available to carry out foot trimming.

Lice can also be a problem in winter, and it is worth applying louse powder or other pesticide every month and clipping out the top of the neck, withers and along the back.

It can be seen that for a successful breeding programme to operate, considerable time, effort and skill are required from herdsmen. As herd health and nutrition are closely involved in fertility, team work is required, so herdsmen need to work effectively with veterinary surgeons and feed advisers as well as with herd owners.

CHAPTER 12

PREPARING STOCK FOR SHOWING AND SALE

Many herdsmen will not be involved in showing or in the need to prepare cattle for sale as the majority of animals culled from the beef or dairy farm go directly to slaughter. Some, however, will be involved in this activity and others may have ambitions to work at a later date with a show herd. Pedigree breeders who participate in shows are provided with a shop window to exhibit the type of animals they may have for sale during the year or in the future. Showing also provides a good opportunity for herdsmen to obtain additional job satisfaction by demonstrating their skills in the preparation and the showing of the animals. Producers wishing to purchase breeding stock should primarily be concerned with production records but, as explained in Chapter 11, the body conformation and appearance of stock is also an important consideration. This chapter is written for those doing the showing rather than for those wishing to buy stock. However, the latter group may wish to read on–despite the fact that it is not the intention to declare too many 'tricks of the trade'!

The first and most important factor in preparing cattle for show or sale is that the job must always be done to a high standard. It is a poor advertisement for a herd if the cattle are only half-washed or are badly clipped. Such animals would make a better price in a sale if left untouched. This is another example of a job which, if it is to be done effectively, must be carried out by a skilled and experienced person who is allowed adequate time and has the required equipment to hand.

Selection of Stock

The process of successful showing starts with effective planning. It is first necessary to obtain the date of the event (perhaps even more than a year in advance) so that action can be taken to have the cattle at an appropriate productive stage for the various classes, e.g. cow in milk. It is also important when, for example, one is selling surplus in-calf heifers, to locate a sale date to coincide with the cattle being at their best, i.e. a few weeks pre-calving.

The animals one intends to enter for a show may have to be selected out from the herd several months before the event so as to enable feeding and management to produce ideal body condition. When halter training is involved this cannot be undertaken with confidence in a short period of time.

Feeding

The feeding of show cattle is an important aspect of preparation and on many farms involves diets somewhat different from those fed to the commercial herd. The main objective is to produce a good bloom to the coat and an adequate level of fleshing (condition scores not usually quoted!). The forage:concentrate ratio will usually be similar to normal diets but it is the constituents of the diet that tend to be different. Bran and sugar beet pulp are ideal foods, helping to stimulate appetite and keep the gut in good function. Oil-rich feeds such as linseed meal are frequently used to help produce the required bright skin and hair. High-quality hay is the most suitable background feed and is convenient to handle, especially at the show. A number of good bales should therefore be put on one side specifically for the show cows.

Managing Show Cows

As most of the shows take place in the summer months animals brought into the buildings for preparation need a cool location with good ventilation. Adequate supplies of clean bedding are obviously required, sawdust being a very useful material. Cattle being prepared for sale during the winter months benefit from being yarded in a separate group to allow appropriate feeding to take place.

Regular exercise is a necessity for show animals, so that time needs to be available to walk the animals, especially a bull, around the farmyard. Even when exercising, it is preferable that a leather rather than a rope halter is used, to avoid rubbing off any hair. The exercise period is a good time to teach the animals to stand well. It involves keeping the fore-legs apart and one

Clipping the udder.

Washing for the show ring.

back leg slightly behind the other. The head should always be held up.

Clipping

This is a very important part of show preparation and needs to be carried out with skill and care. Clipping will need to be undertaken some time before the show or sale, with additional work usually necessary a few days before the event. Electrically driven clippers are most convenient; there is a wide range of cow blades and plates available, including special fine ones for show work. A good grooming before starting the clip prevents the blades losing their edge. Special aerosol sprays are available which, as well as lubricating the blades, disinfect them and also blow out the dirt and hair.

The areas for clipping include the head, neck, shoulders and udder, as well as the tail and topline. One of the skills is to use the technique to minimise defects, so that along the topline, for example, close clipping of high points and a less severe clipping of lower points really does improve appearance. The amount of hair removed at the front or rear of the udder will depend upon the quality of attachment and prominence of the milk vein. Care should always be taken to ensure that the clipped areas blend in well with the unclipped ones, this being easier in summer when there is less hair to deal with. Clipping only the head, top of the shoulder and the tail of cattle for sale can markedly improve their appearance. It is often tempting to clip out the legs as well but this can be a rather dangerous activity when undertaken on young heifers and even on cows!

Washing

This is a routine part of preparation, and preceding the event will need to be done at least weekly. Not only does a good washing remove dirt and dead hair, but it stimulates the skin and generally improves coat condition. Warm water and liquid detergent are ideal (although proprietary shampoos are available), taking care not to allow soap and water into the eyes and ears. Blow-dryers are frequently used after washing to give an improved appearance to the coat. Although stained skin will be rare in show cattle, it can be a problem when preparing animals for sale during the winter months, especially with cubicle refusers–which are the ones that need to be sold! It may be necessary to wash the stained areas on several occasions and then to use some chalk whitening on the day of the sale.

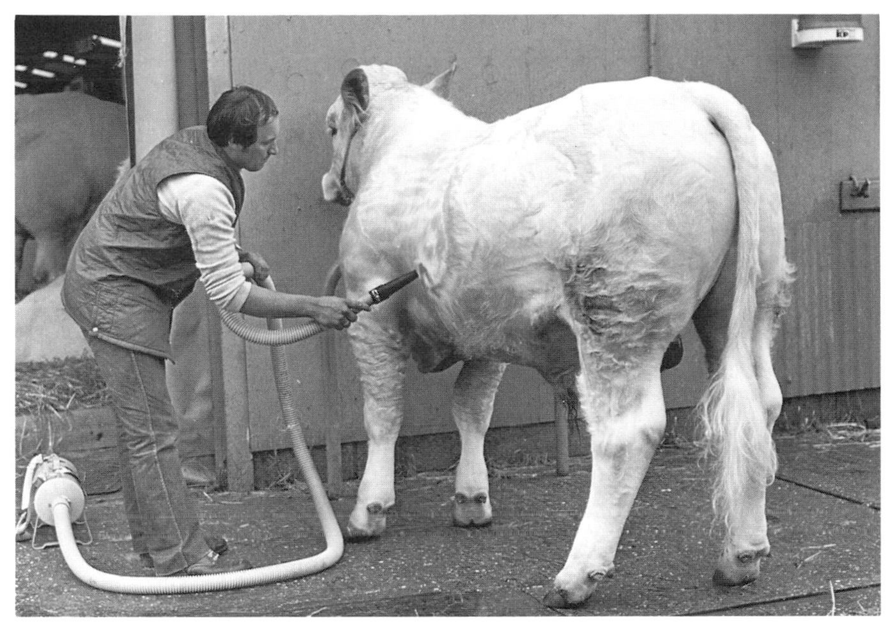

Blow-drying. This is a useful way to improve the appearance of beef cattle with long hair.

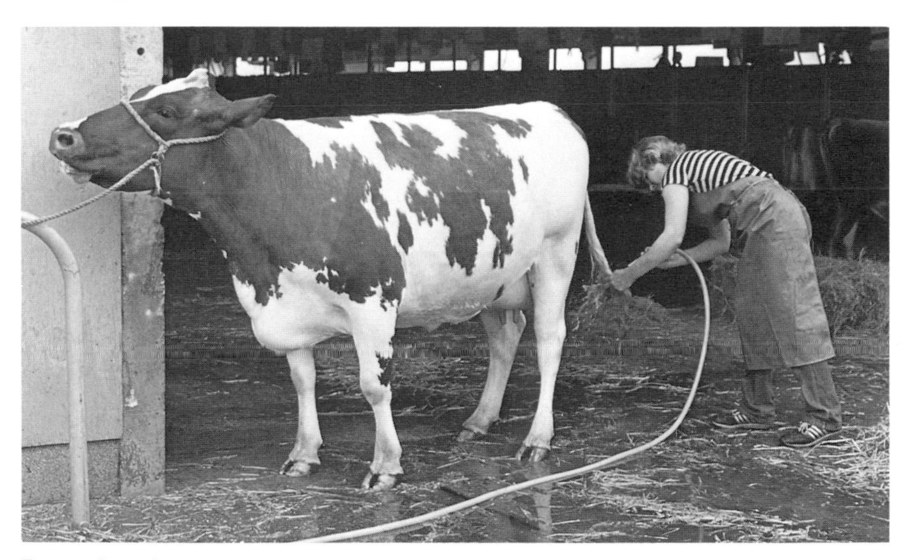

Preparing the tail.

Grooming

After thorough washing and allowing time for the coat to dry, grooming can take place, as ideally it should every day before a show. A stiff brush is best suited for the initial grooming, as is a rubber curry comb, followed by use of a softer brush in conjunction with a lightly oiled cloth, just before parading. The tail needs particular attention, as a well-groomed and fluffed-out tail certainly adds to appearance. The herdsman leading the animal should always be well 'groomed', wearing a clean and well-ironed white coat and polished shoes and having a general neat appearance. Leather show halters and leads also need to be regularly cleaned and polished, a special one being kept for use in the show ring.

Trimming

The feet of show and sale cattle must be in good order. Any paring or cutting of the hoof which is required needs to be done well before the event to allow the animal time to adjust its standing and walking to the new angle of the toes. The horn of the foot looks better if scraped until smooth, followed by polishing with horn or shoe polish. The same technique can be used to trim and polish horns, especially of bulls.

Bagging-up In-milk Cows

It is obviously unwise to milk out a cow several hours before she is to enter the ring, but it is equally important that her udder is not too full, causing restlessness, or that the teats are not beginning to 'poke'. Knowing the time of parading, it is necessary, therefore, to milk out as appropriate, this being another example of the skill of the herdsman in knowing the precise requirements of each cow, and even of each quarter.

A sharp appetite needs to be maintained as the show approaches, with a good feed and drink being allowed just before parading takes place. The likes and dislikes of individual cows should also be known so that a 'little bit of what she fancies' can be offered at the appropriate time.

Parading

There is also some skill in walking the animal slowly but effectively in front of the judge or the buying audience. One needs to be calm but firm with the animals, sometimes walking backwards, so as to be in a position to check that the animal looks at her best. When asked to stand for close inspection by the judge, it is important to have the animal standing well, with head up and legs positioned as described above.

Appearance of Show Stand

Although the parade of the animals in the ring in front of the public and other breeders is the main shop window, much potential business is made or lost by the general appearance and tidiness of the exhibitor's stall area. The equipment and feed should be neatly positioned, keeping most of the tools in the show box. Someone should be 'on duty', especially at busy times, not only to keep an eye on the stock but also to answer questions and generally be helpful to potential customers. The availability of a herd brochure may well be justified; or if not, some performance data (neat and well presented) should be on display above the animals. Many show societies positively encourage such an attitude by awarding prizes to the best breeder's stand. Additional factors taken into account when judging these classes are cleanliness of lines, utilisation of space and overall effect for herd and breed.

Showing can be a pleasurable and satisfying task for the herdsman. It involves long hours, hard work and, on most farms, the work is piling up awaiting his return from the show. There is also the added danger of animals contracting diseases, but for most herdsmen it is a very worthwhile activity.

CHAPTER 13

MAINTENANCE OF MACHINERY AND EQUIPMENT

In the past, the herdsman took little interest in farm machinery as his job was seen to be caring for stock and he was content to leave the caring of those 'mechanical gadgets' to others. Furthermore, farm people with a preference for driving tractors have tended to be located on arable holdings and have avoided too much involvement with farm livestock whenever possible. Occasionally, a herdsman may have been seen oiling the wheel of his barrow but only when the squeak was so serious as to disturb his cows!

Gradually, however, the scene is becoming less clear cut. Mechanisation of cattle units has taken place slowly but progressively over recent decades, first, with the introduction of the milking machine, then feed and manure-handling equipment and now with automation and electronics. It is fair to say that a number of today's stockmen do recognise the importance of the machinery and equipment and take a keen interest in its maintenance but there are still many others who do not. With the continuing trend towards increased herd size and a reduced labour force, however, the essential role of reliable machinery cannot be over-stressed. As with the veterinary surgeon and the cows, outside specialist help is often required for major repair work, but the need for this is reduced if the regular stockmen understand the operation of each machine they use and are involved in the routine servicing and adjustments. It is therefore the case that the new generation of herdsmen need to have a much wider range of skills at their fingertips, although there are limits and it is not suggested that they should be able to write computer programmes or even replace the clutches of tractors! This chapter is intended to help the stockman recognise the value of efficiently maintained equipment by showing that it responds just as much to his care and attention as do his cattle. As with animals, careful observation of machinery is needed and action must be taken at the first signs of a fault.

One of the major problems facing a stockman is to find the time for all the numerous jobs that need to be done each day. Problems seldom appear at a steady rate so that they can be tackled there and then; more often than not one problem leads to another and, of course, the animals must come first. But unless machinery problems are given some regular attention, a vicious circle can be established. Routine jobs undertaken with inefficient machinery take longer than they should so that, in turn, there is less time for vital maintenance. One of the most common examples of this is the bent or damaged slurry scraper which continues to be used, causing additional damage to doorways and even to cubicle uprights. If planning the use of one's time is important, even more so is making sure that those plans are carried out—and especially that the difficult or unpleasant jobs are not just left undone.

Herd owners and managers have a particular responsibility in this area, making sure that they recognise signs of stress in herdsmen, avoiding their constant overloading and arranging for appropriate back-up services. Fewer problems appear to exist on farms where owners and managers are actually involved in the routine tasks, perhaps by undertaking relief duties. This gives them a clear picture of what the job entails, how well the machinery operates and, for instance, whether there are the necessary tools to hand to cope with essential repairs. Small farms and one-man dairy units often tend to suffer most from a lack of adequate time to deal with machinery problems, whereas the larger units usually have at least one member of the team of stockmen with the time and interest required.

Some dairy and beef units are part of much larger farming businesses often associated with arable enterprises so that a farm workshop is available, with staff to deal with the herdsman's equipment problems. Relationships between herdsmen and farm mechanics are not always good, due to their contrasting interests and to a failure of each to understand the other's point of view. Herdsmen can make a positive step to improving the relationship by, for example, making attempts to wash off as much slurry as possible from the damaged scraper before dumping it on the workshop floor. Inviting the mechanic to inspect a problem at an early stage can be far more rewarding to all concerned than waiting for a complete stoppage and then 'passing the buck' to the workshop. Similar attitudes are known to exist in enterprises which are sufficiently large to justify stand-by equipment. If stockmen abuse such a resource then they often find that both

Washing off the slurry scraper for a maintenance check.

Using a power washer to clean equipment.

machines are out of commission and considerable inconvenience occurs.

Even on farms where there is a well-equipped workshop, it is advisable for the cattle unit to have a small set of essential tools for use when the workshop is closed. This should include such items as hammer, pliers and screwdrivers, as well as adjustable spanners and any specialised tools required to repair drive chains. Where more than one person uses the tools, there should be an established storage place where each item can be returned after use. The location would usually be in the dairy office or store, along with such things as calving ropes and medicines. A more comprehensive set of tools will be required on farms without a specialised workshop and these may be so valuable as to justify a lockable store.

The availability of spare parts is another important factor in efficient machinery operation. Carrying too many parts on the farm is expensive because it ties up cash, but carrying too few creates obvious problems with frequent journeys to and from the suppliers. When a new item of equipment is obtained the dealer's experience should be used to help decide which and how many spare parts should be held at the farm. A very satisfactory system for replacing used parts is to attach a label to each important part in the store recording which machine it is for, part number and local supplier. When the part is used, the label is removed and placed in a box marked 'shopping orders'. When someone from the farm is next going to town, they take the labels for part numbers. On return, with the replacement, the same label is attached and the item put into store. A similar well-organised method of storing instruction manuals and parts books is recommended. A small metal cabinet with drawers is ideal and, again, can be located in the 'stores'.

On some farms, the 'boss' may prefer to keep such books in the farmhouse or general office, so in those cases, stockmen should be supplied with copies.

Mention has already been made of the arrival of new items; this is an important occasion, as efficient operation is so often determined by the success of the 'induction ceremony'. All expected users should be present, including 'the Management', to have potential problems pointed out and to agree a maintenance programme. This may not be required for the infamous slurry scraper, but it is essential for such items as bulk tank condensers, forage boxes or electronic feeders. The name and telephone number of the service agent needs to be recorded in the dairy

A slurry separator. A good example of a machine which, if well designed, constructed and maintained, will reduce the time involved by herdsmen in dealing with slurry. If not, breakdowns lead to increased time involvement and frustration.

as well as in the farm office (along with those of the veterinary surgeon and AI centre). It can be advantageous, especially with less common equipment such as a complete diet feeder, to have an arrangement with a neighbouring farm operating a similar item, so that a stock of spares can be shared and information exchanged on solutions to potential problems.

As with cattle breeding records it is a considerable help in the planning and carrying out of machinery maintenance to have some suitable method of recording. A small blackboard or calendar may be adequate, listing the machines and date of the next service. If anyone then has a spare hour during the week, he can check the list and find which item is in most need of attention. Knowing the ways many stockmen keep such records, it may be easier for them to have a few specially coloured pins in the actual breeding board—not indicating when the old tractor is due to calve but when it should have an oil change!

A useful alternative is to use Action Plan Lists to remind one of the jobs to be done before each seasonal activity. Such lists can be obtained from machinery or fertiliser manufacturers, but can be easily prepared. One for use before the winter feeding period would include such items as:

Machinery

Slurry scraper

☐ Check condition of frame, especially wings and rubber blade.
☐ Have spare blades ready.

Forage box

☐ Service according to instruction manual.
☐ Check universal joints, especially at drawbar.
☐ Check spare parts stock, especially drive chains.

All these 'on-farm' aids such as a good tool set and appropriate recording systems considerably help the efficiency of machinery maintenance. But another factor of major importance is the basic understanding of the subject by the person carrying out the task. Many younger stockmen now coming into farming have studied various aspects of mechanisation as part of a general agricultural course. Others may not have have had such an opportunity, so it is to their benefit to identify and take part in specialised training events to improve their knowledge and understanding. The ATB and many agricultural colleges offer such courses and hopefully will increase this activity in the future.

Efficiently operated machines not only cost less in repairs but on the cattle unit they contribute considerably to increased output. Stockmen who can carry out their routine duties effectively and without the worry of frequent breakdowns have more time for their key job, which is to observe and care for their animals.

CHAPTER 14

FIELD WORK

As with the maintenance of machinery discussed in the last chapter, the extent of the involvement of a herdsman in undertaking various non-cattle tasks in the fields will be dependent upon a number of factors including size and type of farm, age, experience and level of training. On the all-grass family farm, everyone involved will usually be proficient, or be prepared to have a go, at most 'outside' jobs, including fencing, silage making and even ploughing. At the other extreme, on the large farm or estate which employs specialist tractor drivers, the herdsman, apart from perhaps moving an electric fence, stays very much with his cattle and in the winter months may not even be involved in feeding them.

The majority of herdsmen, however, welcome an opportunity, when time and routine duties permit, to get out of the buildings into the fresh air to undertake, or at least help with, some field work. As discussed in earlier chapters, most herdsmen recognise the need for timeliness in, for example, the provision of high-quality silage or clean dry straw, so anything they can do to assist with such operations will be to the good of 'their' enterprise. On the larger holdings, there is much that stockmen can do to prevent a 'them and us' attitude, as outlined in the last chapter in respect of the staff of the farm workshop.

Showing an interest by a willingness to make time, especially at busy periods such as harvest, to top a few paddocks or spread some fertiliser can do much for overall 'team spirit'. Such a willing attitude can also help the motivation and, therefore, the cost-effectiveness of contractors if and when they are employed.

Whether the situation demands it, or whether it is just preferable for the stockman to undertake field work, the results of the actual task will be greatly influenced by the input of skills, experience and confidence, as well as by the equipment available and the time allowed. As with cattle husbandry tasks, attendance at appropriate training courses, working with and learning from skilled operators and getting as much practice as possible, all influence output and quality of work. Before undertaking the job it will usually be necessary to check whether any materials required have been ordered and in fact delivered, as well as the availability of suitable equipment.

A number of particular field tasks will now be outlined. The list will not be comprehensive for all situations (e.g. no dry stone walling), nor will it be possible or practical to attempt to describe in full detail how each job could or should be undertaken.

Electric Fencing

The need for, and some explanation of the task of erecting and moving an electric fence, was described in Chapter 4. The various items of equipment such as wire, reel, posts, unit and batteries need to be collected from a specific storage location, with surplus items replaced on return. Metal posts left lying around the farm are a potential hazard, especially to forage harvesters, as are small bits of wire to the cattle. Appropriate transport is advisable because attempting to carry the equipment over any distance can be time consuming and painful!

Having decided on the appropriate location for the fence, the first task is to anchor the reel wrapped with adequate wire, to a firm base, e.g. a special 'corner' post or to the permanent field fence, leaving the reel in the unlocked position so that the wire can be unwrapped as required. For most lengths, it should then be possible to carry all or most of the intermediate posts along the proposed line, putting them off at their approximate position, loosely into the ground (and not lost in the grass), whilst at the same time unwrapping the wire. At the far end, the wire is firmly connected (via an effective insulator) to another anchor point. Before finalising the position of the intermediate posts at some 10–15 m apart, it will be necessary to return to the reel to tighten the wire and activate the locking device.

A straight fenceline is not essential, but if the wire is to be reasonably tight, there will be little alternative and during, as well as after, grazing the result will be available for widespread observation, perhaps from the neighbours! Having connected the unit to the line, put a good 'earthing' post into the soil and switched on, the herdsman needs to check whether the current is satisfactory. Some are 'happy' just to grab the wire; others (along with the author) prefer to hold a piece of grass to the wire. Mains operated units are ideal if the area to be fenced is conveniently located near a building and the wire does not have to cross a public road. This particular task has been described at some length to illustrate that there is considerable detail which needs to be carried out if the job is to be done well and give satisfaction.

Permanent Fencing

This task can involve anything from minor repairs to an existing fence and erection of semi-permanent paddock fences, to a new and elaborate post and rail boundary to the farmyard. As with many farm tasks, it will be possible to operate single-handed, given adequate equipment and time. Fencing is, however, one of those jobs more efficiently undertaken by a team of at least two, particularly for parts of the task such as post driving.

When repairing an old fence, it is helpful if agreement can be reached with the 'powers that be' as to the extent of spending. So often such a job is 'throwing good money after bad', but a stock-proof fence will need to be available, at least until the end of the grazing season.

Semi-permanent fences commonly used for paddock grazing can also utilise a single electrified and appropriately insulated wire but, more often, two strands of barbed wire at 600 and 800 mm above the ground.

For permanent stock-proof boundaries, even with stone walls or a thick hedgerow, it is helpful to have a single strand of wire on the inside (say 1 m distance) to prevent damage by the stock, as well as providing a narrow 'wild-life' strip.

Before erecting a new post and wire fence it is usually necessary to clear away the old one, making sure to retrieve all the materials, i.e. not leaving any old staples to get into the stomach of a cow. Levelling off the ground with a tractor loader bucket may be helpful, so allowing the new wire to follow more effectively the overall contours of the line. The key to success in erecting such a fence is to have very firm corner posts which will not move when full tension is put on the wire. It will pay to have these up to 200 × 200 mm with a length at least 1,500 mm so that 500 mm can be in the ground. They should have a strut fitted in-line with the direction of the wire(s) to give added stability. Such a straining post with struts is also required every 50 metres along the fence. Intermediate posts will be required every 2 m and can be either round with a diameter of 80–100 mm or 75 × 75 mm square. Ideally, a mechanical/hydraulic machine is available to knock the posts into the ground without damaging the top. If a drive-all or mallet has to be used, an assistant is essential. There is considerable skill in keeping the posts in line and also vertical—this also being a two-man job.

When fitting the wire, especially if square section 'pig type' is being used, it is preferable to roll out the full length, firmly connect one end (at the appropriate angle, normally vertical) and then tension the full roll. Special ratchets are available which, when anchored to a tree or tractor, can provide the optimum tension without undue stretching of the wire. When fastening the wire to the intermediate posts, the staples should not be driven in too much so as to damage the wire; this also makes it easier to remove them subsequently should that be required. The use of a marker to indicate height above ground is helpful, especially when attaching barbed wire (heavy leather gloves also required).

Grass Harrowing

Most grazing areas, but especially permanent pasture, benefit from a good harrowing in the later winter/early spring to pull out any dead material. On very stony soils, the practice may not be advisable, as the harrows could pull up stones which could later damage cattle feet. The task of passing over all the field offers a good opportunity to check for (and pick up) any foreign bodies such as bricks or pieces of wire, which have a habit of getting into the middle of pastures!

This is a relatively unskilled task, perhaps best suited to the junior team member or the one who most enjoys tractor driving. It should not however, be undertaken at 'Grand Prix' pace; otherwise the operation will be ineffective and the harrows soon in need of replacement.

As mentioned in respect of other dairy equipment such as calving ropes, when the job is completed, the harrows need to be stored in a suitable location. Leaving them to 'grow' into the grass can be highly dangerous to subsequent passers-by—cattle, people or vehicles.

Rolling

Young grass leys usually benefit from a pass with the flat roller in the spring, so long as the soil conditions are appropriate. If too

A straining post with struts. Extra stability is obtained if one such as this is located every 50 metres along the line.

Tightening barbed wire with a special rachet. Note the use of heavy leather gloves.

wet, the tractor wheels will make unwanted ruts and if too dry the job is a waste of time.

It is recommended that all grassland which is planned to be cut for silage, or even hay, should be rolled not just to firm the plant roots, but as with harrowing, to provide an opportunity to inspect for unwanted foreign bodies. It is helpful to fit a box to the drawbar of the roller as a carrier for such collected items. The cost of repairs and associated delays to the progress of a forage harvester can be markedly reduced if this job is done with care—it may even justify a bonus! As with harrowing, a slow steady speed is required, because repairs to the broken axle of a roller are expensive. Particular care when operating with the roller or harrows needs to be taken on the first pass around the headland if the boundary is a post and wire fence. Much damage can be done to the fence if accidentally caught by the implement.

Pasture Topping

Several times during the grazing season it may be necessary to follow the grazing of an area with the chopping of uneaten material. This is particularly relevant in respect of weeds, e.g. thistles, but also to improve the quality (and appearance—which the author considers important) of the subsequent growth. Modern topping machines usually operate with a rotating steel blade, which in some models can be fitted with replacement sections. Before use it is important to check the blade for damage and to operate slowly 'out of work' to check that the blade is balanced and not causing vibration. Most models have specific areas requiring grease or oil. All have a gear-box with oil levels requiring a routine check and, of course, all must have effective PTO guards. As with other equipment connected by the three-point linkage, care and safety are essential when connecting to the tractor, using the appropriate category pins.

Care should be taken when travelling to the field, and in not lifting the chopper so high as to damage the PTO shaft. In work, ensure an optimum height—to remove unwanted herbage but not to scalp the sward. It is important to look out for any obstacles such as tree stumps and manhole covers, and adjust forward speed using the gears, whilst maintaining the recommended PTO speed. After use, the machine should be cleaned and, if necessary, the topper washed off so as to check for any problems which can then be corrected before it is next required.

Fertiliser Spreading

As this is one of those jobs which needs to be done at the right time, it is preferable for herdsmen to be able to undertake spreading fertiliser if other members of staff are unavailable.

Once the spreader and PTO shaft are connected to the tractor, it is necessary to grease-up or oil as required, then test for effective operation before filling the hopper. Usually the machine will have been previously calibrated for spreading the product, if not, the instruction book will be needed. It is helpful to be provided with the appropriate settings and forward speed for specific application rates, ideally on a record card (from the office)—based on the experience of previous operations. After the job, the actual usage and settings should be recorded on the card before returning it to the office for updating of the individual field records.

It is also of help to know the width of spread for the product being used, so that on arrival in the field one can mark out the centre of alternative bouts at opposite ends. This can be done, for example, with paper bags or brightly coloured string, providing they are visible from the opposite end of the field.

Whilst operating with the recommended PTO speed, it is necessary to select the appropriate gear for the required ground speed and drive as straight as possible towards the 'correct' marker. If there is any uncertainty in respect of calibration, it will be wise to put only a small known quantity (say 2 bags) into the hopper and then after covering a known area (a paddock of 1 ha) make the necessary adjustments before continuing with the field.

Having completed the job, a thorough washing of the spreader will be required. Even after a temporary halt of only a few days, perhaps to allow the grazing of the next paddock to be completed, it is as well to wash in case of unexpected delays. Storage on level ground at an appropriate height, by, for instance, using a pallet, makes reconnection easier next time.

Manure and Slurry Spreading

This is a task which causes few problems on farms with well-maintained equipment, motivated staff and sound management practices, but disasters on others. A wide range of machinery and equipment is available for the storage, transport and spreading of solid, semi-solid and liquid materials. Problems are minimised if the appropriate machinery is available to undertake a specific operation—solved by many in employing contractors to undertake specific tasks, e.g. emptying and spreading material from a lagoon or agitating an above-ground store which has 'crusted' over.

Herdsmen have an important role to play in ensuring that only the appropriate materials get into storage and handling systems. Straw and waste feed, for example, need to be kept out of liquid systems, as do strings, bricks and pieces of wood from solid systems (or in fact from any system).

Trailed spreaders and tankers need to be operated with a tractor of the appropriate size, particular care being taken on hilly sites and especially when turning. It is essential to avoid overfilling open-topped spreaders, especially when travelling on public roads, as not only is spillage a hazard but it encourages the general public to think that farmers have little care for the resources they manage.

When spreading, there is often the temptation (due to lack of time and pressure of other work) to 'get rid of' the load quickly and as near the gate as possible. However, if good grass crop growth (or any other crop) is the aim, and the cost of its production is to be controlled, then the optimum use of the nutrients in the slurry and manure needs to be obtained. Although it is not possible to get a particularly uniform spread with some spreaders, it is worthwhile at least to try to cover the land as evenly as possible. Regulations in respect of spreading will no doubt tighten in the future so that such aspects as discussed above will become even more critical.

Tankers and spreaders in regular use obviously do not justify washing-off after each operation, but it is vital that at regular intervals a thorough wash (ideally using a power wash) is undertaken so as to check for any wear or mechanical problems. Some time may have to be spent cutting strings away from the shaft of a rotating-flail type machine—which brings us back to where this subject was introduced: such strings should not have been allowed to get into the system in the first place!

Silage Making

A separate chapter, or even a book, could be written on this subject—and some have. The role of the herdsman in *assisting* with the various tasks will therefore be concentrated upon. Preparation of the clamps and fitting of any plastic side sheeting is assumed to have already been undertaken.

The need for the flat roller to go out weeks ahead has already been discussed above. Mowing for wilting will usually be the first task of the campaign and one which in several respects can conveniently be undertaken by herdsmen. So long as sufficient grass is wilting ahead of the forager, cutting can operate independent of the other operations, i.e. it can start and stop to fit in with other duties such as milking or perhaps a calving. In order to achieve optimum sugar levels in the herbage, cutting during the afternoon and evening is preferable to early morning—which should fit in well with the herdsman's availability.

Driving and maintaining a modern sophisticated forage harvester is a very specialised task and, if to be undertaken effectively, involves appropriate experience and training. Herdsmen can gain such experience by perhaps taking over for the occasional load under the supervision of the head tractor driver. Accurate driving is essential so as to pick up all the swath together with careful adjustment of the spout, to ensure that as much material as possible arrives in the trailer and not back on the ground.

When additive is being applied, it is necessary to ensure that the optimum quantity is dispensed and that all safety measures are undertaken. The herdsman may find a useful role in delivering supplies of additive to the field or perhaps assisting, as necessary, with the changeover of large drums back in the yard.

Assisting with transport is a more usual task for herdsmen, especially when distant fields are being harvested, as the use of an additional trailer assists in keeping the harvester on the move. If side-loading is practised, close observation of the movement of the forager has to be maintained, and if trailers are pulled behind the forager, parking the empty one in the best position for efficient changeover is particularly helpful. Delivering the full load to the clamp is not difficult on good farm roads but can be a considerable challenge along uneven, pot-holed tracks. (Road repairs need to be discussed with the boss before the next cut!) Travelling along public roads has potential problems, particularly when turning right—or even left if it is necessary to swing out to get square for a narrow gateway. Fully operational indicators are a 'must' for road work, their location being critical to avoid damage when emptying. Effective communication with the buckrake or loader driver is needed to ensure the tipping of loads into the required position. The availability of trailers with automatic tail-gate opening is a great boon to speed of turn-around.

Herdsmen tend also to be regularly involved on the clamp as loader drivers, or perhaps just assisting with consolidation by extra rolling. This is one of the key jobs in the whole campaign, as effective filling has a considerable influence on herbage fermentation, thus affecting intake characteristics of the silage. It is important to load evenly, to concentrate particularly and compact well alongside the clamp edges and, as the height builds up, to take great care not to drive too close to the safety rail.

Overnight coverage of the clamp is a job often avoided but if quality silage is the aim, it is well justified. The willing assistance of herdsmen in this task, even if they are not involved in any other parts of the operation, will be greatly appreciated by the others. Following completion of clamp filling, a quick but effective sealing needs to be undertaken—usually using car tyres or bales to hold down the plastic sheeting. This definitely is a team job and on completion, the members deserve to have an early bath, re-assembling in the local 'hostel' for the silage equivalent of a harvest supper—and it should be 'on the house'!

Big bale silage, bagged or preferably wrapped, is increasing in popularity on many farms, if only for use in buffer feeding

ACTION PLAN FOR SILAGE MAKING

MACHINERY
Mower/Harvester

- ☐ Check condition of all blades and the clearance between concave and stationary blades.
- ☐ Have spare blades ready.
- ☐ Lubricate and check free play on chains.
- ☐ Lubricate gearbox.
- ☐ Check universal joints, especially at drawbar.
- ☐ Check tightness of all nuts and bolts.
- ☐ Check bearings for wear.

TRAILERS

- ☐ Check condition of all tyres.
- ☐ Check condition of ram seals.
- ☐ Look for cracks in the body, subframe, axle(s) and drawbar.

ADDITIVE APPLICATOR

- ☐ Check condition of all hoses and jubilee clips.
- ☐ Look for cracks in plastic parts, e.g. can fitting.
- ☐ Replace rubber parts.
- ☐ Have a spares kit ready.
- ☐ Inspect condition of any electrical leads and contacts.
- ☐ Do not refill cans with additive, especially on pressure applicators.
- ☐ Check for leaks using water.

ALL MACHINERY

- ☐ Book your contractor (if required) now.
- ☐ Service all machinery according to the instruction manual.
- ☐ Check all machinery for loose bolts which might fall off and damage the forage harvester.

Figure 13. If responsibility has to be undertaken for the whole silage making campaign, it is helpful to prepare an action plan, ticking off each item when completed.

programmes. Herdsmen, who may be required to assist in the bagging and transport, need to ensure a good, level and protected site for storage.

Reseeding

The provision of new leys on a farm with an arable rotation can be a routine matter, but on an all-grass holding, especially with permanent grass, the feasibility of a reseeding programme has to be questioned. The high cost of seed and cultivations, as well as the time out of production, needs to be put against the higher yield of grass in the early years following the reseed. If and when a programme is agreed, careful planning is required, taking into account season, weather and how the new ley is to be utilised. Soil tests should be undertaken for pH, phosphate and potash, and deficiencies corrected with the appropriate applications of lime and fertiliser.

The ideal time for a direct reseed is late summer to early autumn, so that a light grazing can take place before the winter months. Burning off the old sward with herbicide (see below on safe spraying) is worthy of consideration, particularly if there is a high percentage of weed grasses such as annual meadow grass. Ploughing is then the next operation, although in some situations, e.g. very shallow soils, a rotovator can be used. Ploughing is another skill in which few herdsmen will be expected to be proficient. In many situations it is better undertaken by contractors, as are subsequent seed-bed preparations, so avoiding the capital cost of equipment which is used. The key factor for post-ploughing cultivations is the production of a fine firm and level seed-bed, without pulling up any of the old turf and, in dry conditions, not losing all the moisture. This involves the use of disc harrows in preference to spring-tine cultivators, with harrowing and rolling at alternative angles. Following the application of the seed by drill or spreader (again, optimum calibration being vital), a final harrow and roll should complete the task.

Weed control in a new ley is seldom a problem, if seed germination and subsequent growth are effective, following suitable seed-bed preparation and reasonable weather conditions. Any weeds that do grow are routinely topped following the first grazing. Occasionally, however, growth will be such as to require herbicide application, probably following advice from an agronomist. Again the use of a contractor would be ideal, as the task of spraying is now one of the jobs for which the operator has to have a Certificate of Competence. The FEPA (Food and Environment Protection Act) regulations of 1986 involve a foundation module (minimum age 17 years) which includes carrying out the correct procedures for storage, handling and mixing of pesticides. This is then followed by additional modules for specific machines, e.g. ground crop sprayers with hydraulic nozzles to hand-held applicators for granules.

In conclusion, it can be seen that some herdsman undertake a comprehensive range of field work tasks. Others, understandably, prefer to concentrate their skills and input to the needs of their cattle and the close environs of the buildings. I return, however, to where I began this chapter, with an encouragement to all herdsmen, whatever their role, to cultivate a genuine concern and interest in what goes on beyond the farmyard. The performance of 'their' cattle depends just as much on the timeliness and quality of work in the field as in the parlour or cubicle house.

CHAPTER 15

RECORDS AND RECORDING

Paperwork is not the most pleasurable part of the job for many herdsmen, although most do appreciate the need in a present-day enterprise for some records to be kept. In earlier years, generally with much smaller herds, minimal recording was necessary because the herdsman's memory, aided by a few dates marked on the cowshed wall, served the purpose well. The effective management of a cattle enterprise in the current economic climate, however, requires the collection, analysis and subsequent use of considerable quantities of data. The role of an individual herdsman in this task will vary widely from farm to farm, but whenever it is undertaken it is another important task requiring once again care and attention to detail. Accuracy is most essential as incorrect information is not only a waste of time and paper but can create extra cost and loss of output.

The advantages of an effective system of records should be obvious to the herdsman, as much of the data recorded is subsequently used by him to carry out routine tasks such as drying off or diverting cows for pregnancy diagnosis. Other data he may be expected to record such as feed usage and fertiliser application is required for management purposes with which a good herdsman should be concerned.

Some records such as livestock movements have to be kept by law, but others are entirely optional. One problem is that there is such a wide range of recording systems available, especially for dairy enterprises. Time can be wasted and confusion is frequently caused if several systems are in operation, involving some duplication of recording. There is a considerable responsibility for herd owners and managers to set up a simple but effective system for herdsmen to operate. Everyone working with cattle will at some time be involved in record keeping if only noting the number of a cow seen in oestrus. During weekends or holiday periods such responsibility may be temporarily increased so that a clear understanding of what is involved should be communicated to relief stockmen. A serious problem can arise if essential data is not recorded, as it can if the head stockman's books are incorrectly completed by an over-keen farm student!

Following collection, much of the data has to be analysed (worked upon) before it can be used in decision making, e.g. in culling. The involvement of the herdsman in this analysis will also vary from farm to farm depending largely upon the back-up service provided by the owner, farm office or off-farm organisation. Some of the events, such as calvings, will be written down and passed 'down the line' but they will also be entered on the herdsman's visual display board.

In the few very modern enterprises, data such as individual milk yields is automatically collected and fed into the memory of the on-farm computer. The data is processed immediately so that any cow required for special treatment, e.g. antibiotic milk to be dumped, is identified by a flashing light on her stall. The record-collecting role of the herdsmen on such farms is very much simplified so long as the equipment is reliable.

This chapter is written to aid herdsmen undertake the basic task of keeping and using records, rather than attempting to describe the various alternative systems used within cattle enterprises. Advice on the selection of an appropriate system for a particular enterprise is readily available from such organisations as ADAS, Genus (MMB) and MLC. Several feedstuffs and fertiliser manufacturing companies, as well as offering enterprise costing and monitoring schemes, also supply useful cards and charts on which herdsmen can record raw data.

As an aid to accuracy, timely recording is essential. Events need to be recorded as soon as possible after they take place by being written into a book or on to a card, chalked on to a blackboard or even punched into a computer keyboard. Delays lead to memory failures and time wasted, for example, having to go back to check on the number of a cow seen in oestrus. Keeping a small notebook and pencil always on hand avoids this sort of problem. The ideal type is one which easily fits into an overall pocket and contains the pencil within the spine. Children's magic slates are used on some farms with herdsmen effectively scratching information on with a finger nail! A period of time needs to be built into the daily routine for recording essential events. This is particularly important at busy times such as the peak of calvings when, although the herdsman is busy with the cows, even more data has to be recorded.

Although the importance of good records has been stressed,

there is always the need for a balanced approach to the herdsman's job, and of course the animals should come first. It has been known for some herdsmen (as well as managers) to be so preoccupied with records that they spend excessive periods of time in the office, to the neglect of their stock or, more commonly, to the neglect of appropriate support for their junior colleagues.

Although, as mentioned, not all recording systems will be described, there now follows an explanation of a few key records in which most herdsmen will usually be involved.

Livestock Inventory

It is essential at all times to know the number, class and location of every animal in the enterprise. Such information is required at regular intervals for completing the MAFF returns (surveys of livestock numbers) but is of valuable general use in the day-to-day management of the herd. A simple monthly chart with a line for a day and a column for each class of stock is appropriate.

Livestock Inventory January 1990

	Cows in milk	*Dry cows*	*Incalf heifers*	*Yearlings*	*Heifer calves*	*Bull calves*	*Notes*
1	95	10	12	25	17	0	
2	96	10	11	25	18	0	Heifer 437 calved H. Calf
3	96	10	17	19	18	0	6 Heifers PD + ve

In a large dairy herd subdivided into groups it is helpful to maintain a record of the location of each animal so that time is not wasted when one needs to be checked or has to be diverted for any purpose. A wall-mounted chart serves this purpose with animals listed in numerical order and with different coloured pins for each group stuck into the board, adjacent to the number.

A similar display can be used to locate groups of grazing animals by sticking large-headed pins into the appropriate fields on a wall-mounted farm map. The number of animals in each group can be written on the head of the pin, a system ideal for assisting relief staff involved in checking outlying groups.

Herd Grouping						
Red = High yield (Yard 4) Blue = Medium (Yard 3) Green = Low yield (Yard 2) Yellow = Dry						
1	R	156	Y			
2	B	157	G			
3	R	158	SOLD			
4	G	159	G			
5	Y					

Outputs (Sales)

On a dairy enterprise, herdsmen need to retain the daily milk slips left by the tanker driver. These are handed in to the office where they are checked against the monthly payment statement. Usage of milk by calves, staff and other outlets (retail sales) may also need to be recorded and a special monthly sheet prepared.

Cattle dispatched from the farm for sale, such as calves, culls or beef animals, should be accompanied by a dispatch note, especially when an 'outside' haulier is employed. A triplicate pad is ideal for this purpose allowing the top copy to go to the purchaser with the consignment, second copy to the office for checking against receipts, and the third to be retained in the book with the herdsman. Details should include date, numbers, identification, age and sex, as well as the purchaser. Although it is never a pleasant task to record deaths or sales to knacker's yard, it is essential for this to be done if the books are to balance at the end of the year.

Inputs (Purchases)

Herdsmen should be responsible for receiving deliveries to their enterprises, such as feed, fuel or spare parts. Quantities and other necessary details should be checked and any errors recorded before signing a delivery note. Notes should then be retained, perhaps in a clip on the dairy wall, until forwarding to the office. Incoming sacks or containers which are not on pallets should be stacked neatly, so that subsequently they can be readily counted, perhaps for the end-of-month or annual valuation. Bulk containers (concentrate hoppers, fuel or molasses tanks) should ideally be calibrated, to help in valuation but also to enable stock usage to be more readily monitored. It is also of considerable help if hay and straw barns, as well as silage clamps, hold 'known' quantities.

Breeding and Fertility

As explained in Chapter 11, an effective system of records is essential to a successful breeding programme. The use of a desk diary may be all that is required in a small dairy or beef enterprise to record daily events such as calvings and services. With only a few animals involved it is not difficult to look back, for example, over a three-week period to check on a previous service. Using a set of breeding tables, forward events such as drying-off dates can be calculated for entering in the diary.

In larger herds, a more comprehensive system of breeding records is justified involving some type of visual aid board in

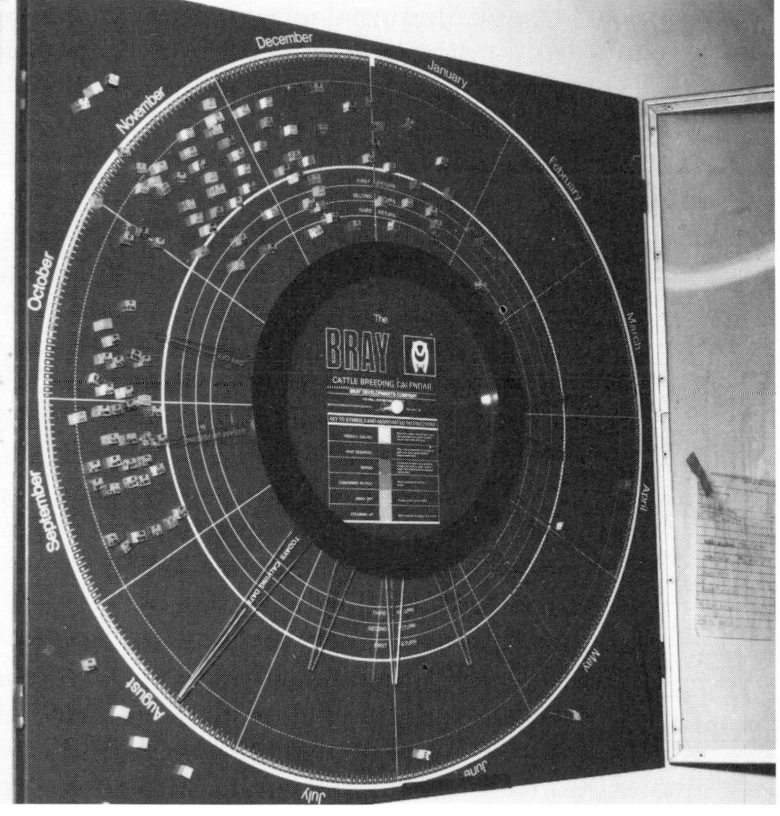

A rotary breeding board. In this example, each cow is represented by a numbered magnetic cube.

addition to written documents. Various types of rotary and oblong boards are available which utilise coloured pins or markers to record specific events. The pointer or marker is moved forward each day, highlighting expected or planned events such as cows due for pregnancy diagnosis. Detailed operating instructions are normally provided by the manufacturers, and it is essential for everyone involved, especially relief staff, to fully understand the system. For very large herds, where two or more such boards would be required to handle all the cows, other systems such as a card index system carrying event markers have been developed. A most satisfactory card was prepared by a commercial company for the 350-cow Sonning herd of the University of Reading. Before computerisation this was used for several years to record in writing the breeding management events on individual cow cards which were appropriately spaced in racks in a cabinet. Small coloured metal slip-on tags were placed on the right-hand edge of each card, which was divided into weekly divisions and used to pinpoint key events, as with a visual aid wall board.

Health Records

All health problems should be recorded as an aid to the management of feeding, breeding and culling, as well as improved herd health. A diary may again be adequate in a small herd but individual cow cards are better for larger herds. The record should start with the calf and be maintained throughout the lifetime of the cow. This is an area in which computerised recording can drastically reduce the time involved in analysing health data, but not in recording the original data.

Mastitis records are particularly valuable as an aid to culling problem cows. Original data is best recorded on a parlour blackboard and then transferred to individual cow cards placed in numerical order in a loose leaf booklet. The information recorded would be:

Cow No:..

Date	*Quarter(s)*	*Tubes*	*Sample Result*

Milk Recording

Official milk recording is organised in the UK by the Milk Marketing Boards. Milk recording can, however, be done privately but membership of an official scheme offers strict standards, a regular discipline and credibility of the results. The yield can be obtained by weighing in bucket plants, and in parlours using graduated jars, milk meters or the new electronic weighers. For the official scheme, a recorder visits the farm on two (or three with 3 × milking) successive milkings each month to record yield and to take a sample for BF and protein analysis. The recorder also takes details of calvings, services, drying-off and cows sold. He leaves a carbon copy of the yields, and the results, incorporating milk quality analysis, are normally received ten to fourteen days later. At the completion of each lactation a record card is received for each cow, detailing her lifetime performance and indicating her relative breeding merit within the herd, and nationally. An annual summary for the herd at the end of March details the performance of each animal and averages for the herd as well as the calving index.

With the cell count of bulk milk supplies soon to be used as another basis of payment, the Board is introducing an optional service in the National Milk Records (NMR) for testing cell counts for individual cows. This will be an invaluable aid to culling animals with ongoing high counts.

Although involved in an official scheme, it is worthwhile undertaking more frequent recording, i.e. every week or two weeks, especially for cows in early lactation. This is a job well suited to the manager or owner, who not only spends valuable time in the parlour monitoring the cows, but is able to talk with stockmen and discuss herd performance.

Monthly Costing Schemes

Numerous schemes such as ADAS Milk Cheque, and Genus (MMB) Herdwatch and Milkminder, are available to dairy farms for budgeting, costing and monitoring herd performance, and feed and fertiliser companies offer a wide range of others. The data to be recorded at the end of the month normally includes milk sales, cow numbers, feed and fertiliser usage. Herdsmen should take a keen interest in the financial as well as the physical performance of 'their' herd and be involved in the interpretation of the results as much as in providing the basic data. Figure 15

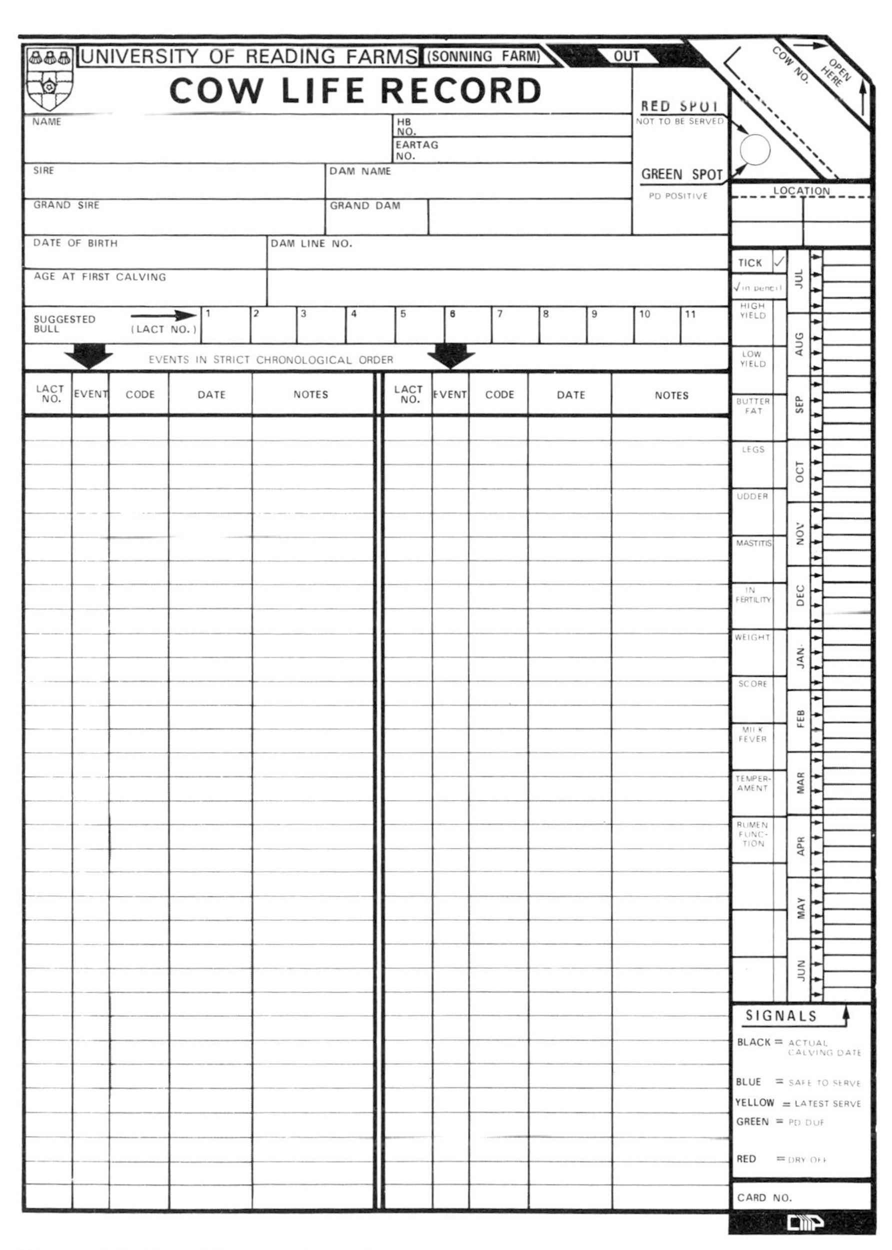

UNIVERSITY OF READING FARMS (SONNING FARM) OUT

COW LIFE RECORD

COW NO. OPEN HERE

RED SPOT NOT TO BE SERVED

GREEN SPOT PD POSITIVE

NAME

HB NO.

EARTAG NO.

SIRE

DAM NAME

GRAND SIRE

GRAND DAM

DATE OF BIRTH

DAM LINE NO.

AGE AT FIRST CALVING

SUGGESTED BULL (LACT NO.) 1 2 3 4 5 6 7 8 9 10 11

EVENTS IN STRICT CHRONOLOGICAL ORDER

LACT NO.	EVENT	CODE	DATE	NOTES	LACT NO.	EVENT	CODE	DATE	NOTES

LOCATION

TICK ✓ (√ in pencil)

HIGH YIELD

LOW YIELD

BUTTER FAT

LEGS

UDDER

MASTITIS

IN FERTILITY

WEIGHT

SCORE

MILK FEVER

TEMPERAMENT

RUMEN FUNCTION

JUL AUG SEP OCT NOV DEC JAN FEB MAR APR MAY JUN

SIGNALS

BLACK = ACTUAL CALVING DATE

BLUE = SAFE TO SERVE

YELLOW = LATEST SERVE

GREEN = PD DUE

RED = DRY OFF

CARD NO.

Figure 14. Cow life record card.

is an example of a monthly results sheet for a farmer using the ADAS Scheme. Performance figures are presented in the upper part of the table for the whole herd, followed by per cow, per litre and per hectare. The first column gives actual results for the month; next is the plan with the positive and negative differences; then data for the same month last year and, in the right-hand column, results for the last (or rolling) twelve-month period. Ideally, following analysis of the results, action is taken to correct deviations from targets. As the information is by this time somewhat dated, it is better to keep a weekly or even daily eye on performance rather than wait for the monthly statement. With many schemes an adviser visits the farm to either collect the data or help with its interpretation. The latter role can be most valuable in using a 'new' pair of eyes to spot problems and discuss (based on experience in the district) the possible alternative actions.

Computerised Systems

Most of the above costing and monitoring schemes use computer facilities, when the data from the farm is normally punched into the system by a member of staff at the bureau or central office of the organisation. Some advisers bring to the farm a mini-computer which can 'stand-alone' or be linked by telephone back to a larger machine, so that data can be analysed immediately. Other schemes involve a group of farmers sharing a facility and perhaps employing a secretary/technician to process the data. An increasing number of farms have their own computer which offers herdsmen more frequent access to the information, should it be required.

With all computerised systems, accuracy of the input data is critical so that a simple but effective system of recording the raw data is essential. A duplicate book is ideal, as used by many farms involved with DAISY, the Dairy Information System of the University of Reading. Events are recorded as they happen, e.g. calvings, services, health problems, with the top copy going to the office on a daily or weekly basis for input into the computer programme. On some farms it is part of the herdsman's duties to use the keyboard and to input the data directly into the system, so that soap, water and a towel will also need to be on hand!

Output from the programme can be in the form of action lists for herdsmen (on paper of easily handled size) or as displays on

Farm number 02.1.0002 Herd 01

		Results	Plan	Difference +	Difference -	April last year	12 month rolling results	
Whole herd	Milk value	**20600.7**	21257		656.3	20659.5	**242806**	£
	Total milk yield	**125683**	130000		4317	126265	**1331694**	ℓ
	Daily milk produced	**4189.4**	4333		143.6	4208.8	**3648**	ℓ
	Concentrates used	**33.48**	30.00	3.48		44.97	**438.5**	*t*
	Concentrate price	**131.2**	140.0		8.8	125.7	**127.7**	£/*t*
	Concentrate cost	**4392.81**	4200	192.81		5650.92	**55981**	£
	Other purchased feed cost	**195.0**	500		305	0.0	**1128**	£
	Margin over all purchased feed	**16012.9**	16557		544.1	15008.6	**185697**	£
	Cows in herd	**199**	199			208	**210**	
	Cows in milk	**195**	195			208	**180**	
	Calvings	**0**	0			0	**194**	
	Dry cows	**4**	4			0	**31**	
	Percent dry cows	**2.0**				0	**14.6**	
	Culls	**0**				1	**45**	
Per cow	Milk value	**103.5**	106.8		3.3	99.3	**1154**	£
	Total milk yield	**631.6**	653		21.4	607.0	**6331**	ℓ
	Daily milk yield per cow in milk	**21.5**	22.2		0.7	20.2	**20.3**	ℓ
	Daily milk yield from forage	**8.3**	7	1.3		4.8	**1993**	ℓ
	Concentrates used	**168.2**	151	17.2		216.2	**2085**	*kg*
	Concentrate cost	**22.1**	21.1	1		27.2	**266**	£
	Margin over concentrates	**81.4**	85.7		4.3	72.2	**888**	£
	Other purchased feed cost	**1.0**	2.5		1.5	0.0	**5**	£
	Margin over all purchased feed	**80.5**	83.2		2.7	72.2	**883**	£
Per litre	Milk price	**16.391**	16.351	0.040		16.362	**18.233**	*p*
	Concentrates used	**0.27**	0.23	0.04		0.36	**0.33**	*kg*
	All purchased feed cost	**3.65**	3.62	0.03		4.48	**4.29**	*p*
	Margin over all purchased feed	**12.74**	12.74			11.89	**13.94**	*p*
	Butterfat	**3.91**	3.87	0.04		3.87	**3.98**	%
	Protein	**3.13**	3.15		0.02	3.15	**3.26**	%
	Lactose	**4.65**	4.65			4.65	**4.63**	%
Per hectare	Stocking rate	**2.19**				2.07	**2.19**	*LU*
	Margin over all purchased feed	**176.3**				149.3	**1931**	£
	Margin over feed and fertilizer	**154.3**				140.3	**1835**	£
	Utilized metabolizable energy	**7.53**				5.84	**77.61**	*GJ*

Figure 15. Milk Cheque monthly results for April 1990.

visual display units (VDUs). No doubt in the near future such facilities will not only be provided in the farm office and milking parlour but in the hospital area for use by the vet and herdsman in health control, and even in the herdsman's home!

CHAPTER 16

CONCLUSIONS

The object of this closing chapter is to summarise briefly the major points raised in the book, and, at the same time, to offer a few *final* thoughts to the reader. Some are indeed afterthoughts but others are general ones, concerning the herdsman's job as a whole, and have therefore intentionally been held over until this concluding chapter.

In the Introduction it was stated that with the continued trend towards mechanisation and automation in dairying, the days of the herdsman, as we once knew him, are perhaps numbered. Having read the preceding chapters, herdsmen should have the feeling that nothing is farther from the truth and that, although the nature of their job may well change, they will continue to have a vital role in profitable milk and beef production. These days, change is accepted as a continuing and inescapable feature of the herdsman's job. In earlier years he had, primarily, to love or at least have a strong aptitude for dealing with cows; today he also needs to have an aptitude for handling machinery, for understanding paperwork and, most importantly, for dealing with other people. It is against this backcloth of 'change' that the final thoughts in this book are presented.

The first *final* thought is simply that 'a successful cattle enterprise involves team effort'. The herdsman's fellow 'players' include, as well as the herd owner and other colleagues, the vet, AI inseminator, feed adviser and electronics engineer. A vital tool of the trade is the herdsman's notebook, which contains the telephone numbers of all his team members in case they are urgently needed 'on the pitch'.

Another important feature of the herdsman's job is its variety, requiring him to have a wide range of skills and giving him little opportunity to complain of monotony. Unlike the immigrant workers employed in the huge dairy units of the western USA, contract milking on shifts around the clock, generally speaking the UK herdsman really has to be a stockman. Although routine tasks such as milking and feeding take up large parts of the day, there is much scope for planning and control, if efficient use of the remaining time is to be obtained. Another important *final* thought, then, is that 'the management of the herdsman's time is a critical factor in the success of his enterprise'. In small herds, especially in one-man units, herdsmen have to manage their own time. Some are naturally better at this than others and there is an opportunity for those who are not so good to correct this weakness. In large units, with several herdsmen employed, there is a considerable responsibility on the management to ensure effective planning of time and coordination of the effort of each team member.

On farms where there is poor use of time, there is a strong tendency for something, or someone, to be frequently neglected with detrimental effects upon profitability. It may even be the cows which suffer, or the herdsman's family or, perhaps most commonly, the care and maintenance of plant, equipment and buildings. A crisis can be expected from time to time on every livestock farm. In the enterprises where staff are well organised, this is 'taken in full stride' by herdsmen, putting in even extra effort, by working longer hours until normality returns. When crises frequently occur, this should be a clear indication that staff cannot cope and that the system needs some major change, which will inevitably involve the staff.

Frequent reference has been made in this book to the need for *timely* action to be taken by herdsmen in carrying out a wide range of their tasks, such as feeding, treating sick animals and keeping records. Although to achieve this, good organisation of time is involved, several other key factors are also involved. Here, therefore, are five general and basic points which, in this writer's opinion, can be crucial to the performance of the herdsman.

1. Adequate knowledge As the level of technology on the farm increases, so do the training needs of the staff. Learning therefore has to be an ongoing process, involving experienced herdsmen as well as new entrants. Despite the new technology, the basic skills of observation and caring for the animals remain the number one priority. The role of skilled staff in helping young herdsmen to gain sound experience cannot be overemphasised.

An afterthought, here, from Chapter 8 on herd health: a valuable training session can be undertaken at the slaughter house when cattle from the herd are being culled, probably due to infertility problems. Arrange for the vet to be present as tutor, to demonstrate the working (and non-working) parts of the cow. You will then be able to discuss health problems with much better understanding.

Knowledge of the science and practice of feeding has been highlighted as a major factor in stockmanship, justifying three chapters in this book. A late 1940s quotation from the 'King of cow feeding', Professor Bobby Boutflour, is still most appropriate to summarise the subject: 'Allow the cow to polish the manger with its tongue a dozen times a day.'

2. Adequate motivation Many herdsmen are happy in their work but others lack interest, feel isolated (psychologically as well as physically) and do not 'get on' with other members of the team. It is hoped that the herd owners and managers who read this book will have a better understanding of the role of their herdsmen and will try a little harder to give encouragement and more effective support, especially at busy times.

On their part, herdsmen should encourage effective communication, try to understand the owner's objectives for his farm and not take a job if not in full agreement with how the farm is expected to be run. An interview for a new position is a preliminary to forming a working partnership and candidates should take the opportunity to find out just what the chances are of that partnership being successful. There are so many different employment situations within the industry, especially in terms of responsibility, that one needs to be entirely satisfied that the job being offered really does match the requirements.

3. Adequate facilities It has been emphasised in several chapters (milking, housing and machinery) that there is rarely any correlation between capital spending and financial performance of the enterprise. Good herdsmen appreciate good facilities but they also use their varied skills to 'make the most' of what is available. Employers, however, have a considerable responsibility if they wish to increase the efficiency of labour use by providing functional buildings and reliable equipment at a cost the enterprise can afford. Some facilities, although a minor cost on the business, can be highly beneficial in improving herdsmen's morale, by demonstrating that management cares. The provision of suitable protective clothing that can be hosed down is such an example; it serves its purpose in protecting the herdsman and reduces the work of his family in washing cow-splattered clothing.

4. Home support If the herdsman is to give of his best to the farm it is advantageous for him to have support and understanding at home. The provision of meals when he is ready, as well as clean, dry clothing, is obvious, but after a hard day's work the comforts of a home rather than a house are invaluable. The well-known saying can be adapted to: 'behind every good herdsman is a supporting wife, mother or landlady'. The practice of young people living-in with the farm family is becoming rare, as is the availability of suitable digs in rural areas. The author is far from keen on what seems to be an increasing practice of single herdsmen living (existing) in sparsely furnished farm cottages with little time to prepare adequate meals, let alone undertake essential shopping. Unfortunately, many of the attempts to set up and operate hostels for young people have not been successful, primarily due to cost. So perhaps one recipe for a successful herdsman is to find a good wife; or a husband for a herdswoman!

5. A life apart from the cows Too many herdsmen (including small dairy farmers) are 'tied to the cow's tail'. Everyone working with cattle needs, and deserves, time off. As mentioned earlier, a number are keen gardeners, obtaining relaxation, despite remaining in the sight and sound of the herd. Interest in the job can be markedly improved by visiting other herds or shows (employers please note!), and also by having time away for hobbies. The younger generation of herdsmen tend to be more mobile and have other interests, so in time this problem may well disappear. Senior herdsmen therefore need to learn the skills of delegation. Relief staff respond to increased responsibility, so long as they know exactly what is expected of them. Communicating and helping juniors to learn from their mistakes are the essential elements of delegation.

A Final Conclusion

If one message should have been clear from this book it is quite simply this: a modern herdsman, with few exceptions, is a very fortunate person. He has a job in which he can take pride and from which he can obtain enormous satisfaction. His efforts, abilities and skills are reflected in the appearance and performance of the herd, which is measurable in the bulk tank, as well as in the herd health and breeding records.

The future promises much; some of his past and present-day skills such as oestrus detection and even milking will no doubt

A day out at the Royal Show.

Preparing for a weekend away from the cows.

become redundant as automation develops further. Sex determination of calves, and even the possibility of the entire herd having identical genetics, will replace current breeding programmes. Developments such as these—and many others—will be of increasing value, as reflecting an electronic and genetic-engineering age, and will assume increasing importance as the needs of consumers change. Those changes will not only be reflected in the demand for milk and dairy products, thus so changing genetic and nutritional criteria, but also in the increasing general concern of the public at large for 'green' issues.

The role of the herdsman in this arena will be invaluable in ensuring that every care is given to the animals and to all the resources on the farm. There is every likelihood that people working with stock, but certainly those taking responsibility, must in the future have Certificates of Competence, involving training as well as practical and written tests. One can imagine that such thoughts will be off-putting to a few, but to the majority of herdsmen, keen to raise the general standards of husbandry in the industry, as well as their own status, they will be welcomed.

However, by the time these developments are perfected and implemented, many of today's herdsmen will be enjoying their well-earned retirement—with a life permanently apart from cows! The majority, if not all of them, will be able to look back with satisfaction, hoping that their successors in the rapidly changing environment will get the same feeling from their work with cattle.

INDEX

Farming Press Books

Listed below are a number of the agricultural and veterinary books published by Farming Press. For more information or a free illustrated book list please contact:

Farming Press Books, 4 Friars Courtyard
30–32 Princes Street, Ipswich IP1 1RJ, United Kingdom
Telephone (0473) 241122

The Principles of Dairy Farming • KEN SLATER

An introduction, setting the husbandry and management techniques of dairy farming in its industry context.

A Veterinary Book for Dairy Farmers • ROGER BLOWEY

Deals with the full range of cattle and calf ailments, with the emphasis on preventive medicine.

Cattle Ailments—Recognition and Treatment • EDDIE STRAITON

An ideal quick reference with 300 photographs and a concise, action-oriented text.

Calving the Cow and Care of the Calf • EDDIE STRAITON

A highly illustrated manual offering practical, commonsense guidance.

Cattle Feeding • JOHN OWEN

A detailed account of the principles and practices of cattle feeding, including optimal diet formulation.

Calf Rearing • THICKETT, MITCHELL, HALLOWS

Covers the housed rearing of calves to twelve weeks, reflecting modern experience in a wide variety of situations.

Farming Press Books is part of the Morgan-Grampian Farming Press group which publishes a range of farming magazines: *Arable Farming, Dairy Farmer, Farming News, Livestock Farming, Pig Farming, What's New in Farming*. For a specimen copy of any of these please contact the address above.

<u>1993</u>

9th April. – ~~Milk fever~~. Calved 8.30pm. Bull Calf.

10th April – Milk Fever.

1994   Milk fever.

22nd June - Calcium Borogluconate
5 hrs before calving. - under skin.
Bull Calf - 3.50 Am.

1st July - 10. Am Calcium Borogluconate
intravenus
- Calciject. P.M.D. - under skin

4.30 pm.
another calcium Borogluconate
intravenus and a booster injection
(stimulant).

2nd July. - got up sometime between
midnight - 7am. very stiff hind
legs.